Cradle of Life

Cradle of Life

The Discovery of Earth's Earliest Fossils

J. WILLIAM SCHOPF

Princeton University Press, Princeton and Oxford

Copyright © 1999 by Princeton University Press
Published by Princeton University Press, 41 William Street,
Princeton, New Jersey 08540
In the United Kingdom: Princeton University Press,
3 Market Place, Woodstock,
Oxfordshire OX20 1SY

The Library of Congress has cataloged the cloth edition of this book as follows

Schopf, J. William, 1941–
Cradle of life : the discovery of earth's earliest fossils /
J. William Schopf.
p. cm.
Includes bibliographical references. (p.   –   ) and index.
ISBN 0-691-00230-4 (cl: alk. paper)
1. Life—Origin.   2. Evolutionary paleobiology.
3. Paleontology—Precambrian.   4. Micropaleontology.   I. Title.
QH325.S384   1999
576.8′3—dc21   98-42443

This book has been composed in Times Roman

The paper used in this publication meets
the minimum requirements of
ANSI/NISO Z39.48-1992 (R1997)
(*Permanence of Paper*)

www.pup.princeton.edu

Printed in the United States of America

3   5   7   9   10   8   6   4   2

To my teachers,
from whom I learned
and my students,
who teach me still

Contents

Prologue

This book chronicles an amazing breakthrough in biologic and geo-logic science—the discovery of a vast, ancient, missing fossil record that extends life's roots to the most remote reaches of the geologic past. At long last, after a century of unrewarded search, the earliest 85% of the history of life on Earth has been uncovered to forever change our understanding of how evolution works.

My own role in the hunt for the ancient life dates from my student days in the 1960s, when active studies were nearly ready to take hold. Apparently the first to prepare at a young age for a career in this field, I have spent that career tracing life's earliest history and have had the privilege and supreme pleasure of seeing this young science sprout, grow, and blossom into a vibrant venture worldwide.

My lifelong involvement in this endeavor has led me to write parts of this book in the first person. For a science book, this is unusual. In the guise of objectivity, we who "do science" usually present our views in a more distant way, often writing in the third person ("it is reported . . . ," "the data indicate . . .") as though the claims made were someone else's, not our own. But I am not objective about this subject—it's my life, I care about it, and it would be false for me to pretend otherwise. Moreover, it seems to me a lot more fun to read about how science is actually done, and by whom and why, rather than plow through a stuffy accounting of theories and facts. "Fun" is the operative word here. To me, science is enormously good fun! There's hardly anything better than learning something brand new or having a novel idea and then following that notion and finding that it makes sense.

So, the goal of this work is to bring to light one of the truly re-markable breakthroughs in the annals of natural science, the discovery

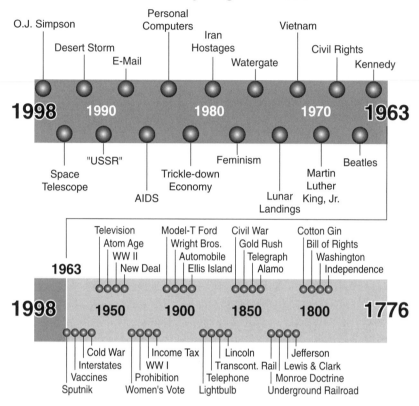

Prologue Figure What if history began in 1963?

of a long-missing fossil record that, by revealing life's earliest history, tells us where we fit in the pattern. And while recounting this story, I also want to show how the science itself evolved—why it took so long for the hidden record to emerge—and convey some flavor of my joy in being part of the endeavor.

A Fable: What If History Began in 1963?

Think for a moment how extraordinary it is that the earliest 85% of life's history has until now remained a mystery. What would it be like if more than four-fifths of America's past were totally unknown?

The year is 1998. The place, a dorm room at UCLA in West Los Angeles. A second-year college student sits at his desk, struggling to cram into his head pivotal facts, dates, and events for his upcoming mid-term in American History. It's good stuff, but he's perplexed—there's so much to learn, all the way back to 1963! President Kennedy's assassination, then Martin Luther King, Jr., then the president's brother Bobby . . . sit-ins, civil rights, Vietnam, flower children . . . space walks, lunar landings, computers, E-mail . . . feminists, AIDS, downfall of the "evil empire." Such a lot to sort out!

Exhausted, he daydreams: What happened before 1963? No one seems to know. The professor once raised the question, explaining only that "a pre-1963 historical record ought to exist— something *must* have happened in earlier decades—but there are no facts to go on. No one knows what happened, or why the record's been wiped out. It's one of history's greatest puzzles."

As the student treks across campus to take his exam, he picks up a copy of *The Daily Bruin*, the student newspaper. Emblazoned in type 3 inches high is the bannner headline: "ANCIENT ARCHIVES DISCOVERED—U.S. DATES FROM 1776!" Excitedly he pours through the article. *"Researchers report that conclusive evidence of the earliest 85% of the history of the United States of America—from 1776 to 1963—has been discovered. Long thought forever lost, new finds document an unknown and unimagined early history of the country . . . a Declaration of Independence from British rule, a written Constitution . . . Washington, Franklin, Jefferson . . . Lincoln, the Roosevelts, a feisty Harry Truman . . . electricity, telephones, radio, television . . . transcontinental railroad, Model T Fords, airplanes, rocket-powered flight . . . Abolition, Prohibition, women earn the right to vote . . . the Dust Bowl, a Great Depression, the United Nations, the Nuclear Age. . . ."*

Astounding! For the first time, hard facts are known that can tell the student how his country began, then grew and prospered over nearly 200 years that seemed lost forever. The traditional history, the post-1963 epoch he learned so well, is only the latest chapter of a very much longer volume!

An even more mind-boggling tale of new discovery unfolds in the pages that follow, but it is scaled in millions and billions of years rather than a mere two centuries, and it deals with all of life, over all of time, over the entire globe. By revealing our roots and unveiling our past, it, too, tells us where we have come from and who we are.

Acknowledgments

In writing this book I have had the good fortune to receive help from many friends, students, and colleagues. Richard Mantonya assisted in the preparation of several of the figures. Those who provided comments and suggested ways to improve the text include John Bragin, Walter Fitch, Henry Gee, Mott Green, Chris House, Amir Lagstein, Xiao Li, Patricia McCarthy, Alice Ormbsy, James C. Schopf, Erik Schultes, Jane Shen-Miller, Karl Stetter, Dawn Sumner, Cindy van Dover, Paul Taylor, and Alice Calaprice, Princeton University Press's gifted senior editor. Greg Stock was particularly generous in his efforts to help me improve my prose. I am grateful to all and hope their suggestions have been properly taken into account.

I thank also Karl Stetter and the staff and students of the Lehrstuhl für Mikrobiologie, Universität Regensburg, Germany, where the final manuscript was prepared during my tenure as an Alexander von Humboldt senior research awardee, and my friends and colleagues of the Precambrian Paleobiology Research Group, several of whom provided illustrations for the volume and all of whom have taught me much over the years. Research results summarized here are from studies supported by NASA Grant NAGW-2147.

Above all, I thank my wife, Jane. She read the text through its many revisions and deserves great credit for offering suggestions time and again that were right on the mark. Her help, and support, have been invaluable.

Cradle of Life

Darwin's Dilemma

Breakthrough to the Ancient Past

Over the last three decades, the evolutionary Tree of Life has been extended sevenfold. An immense early fossil record, unknown and thought unknowable, has been discovered. For the first time we have firm knowledge that life originated, evolved, and rose to become a flourishing success during the infancy of planet Earth. By 3,500 million years ago, a scant 400 million years after the planet had become liveable, life was already well advanced.

Before this breakthrough:

- No one had foreseen that the beginnings of life occurred so astonishingly early.

- No one had guessed that Earth was inhabited only by diverse, vanishingly small forms of life throughout the earliest four-fifths of its existence.

- No one had imagined that the modern world —the familar fauna and flora of air-breathers and oxygen producers, the eaters and the eatees —is merely a scaled-up version of a microbial menagerie billions of years old.

- No one had surmised that evolution itself evolved over geologic time, that the rules of the Darwinian struggle changed dramatically as the history of life unfolded.

This is new knowledge, the result of the last three decades of active research. But the discoveries trace their roots to questions first raised many years ago. Indeed, the dilemma posed by the missing early fossil record—a void in the history of life that to some seemed to under-

mine the foundation of Darwin's evolution—was already widely recognized in the mid-1800s.

The attempt to fill this void has a long acrimonious history of false
starts and embarrassing mistakes. But to understand this saga, we first
need to know about the nature of geologic time and the historical
development of the geologic timescale.

The Nature of Geologic Time

We speak easily, even glibly, about the geologic past in terms of hundreds of millions, even thousands of millions—literally *billions*—of
years. But what do such gigantic numbers mean? Whether applied to
the national debt, stars in the sky (50 billion galaxies in the visible
universe, each 100 billion stars strong), or neurons in the human brain
(more than 100 billion with 100 trillion connections), they are simply
so huge, so astronomical, they are all but incomprehensible. This is
true, too, when we speak of geologic time. But we can come to grips
with time by understanding how it is subdivided—just as a year is
divided into months and days, or a day into hours and minutes—and
then calibrating it by a sequence of events (breakfast, classes, lunch,
schoolwork, dinner, TV, bedtime) for geologic time using events in
the history of life.

The Geologic Timescale

By international agreement, all of geologic time, the total history of
the Earth, is divided into two major "eons," the Precambrian and the
Phanerozoic.

The older, the Precambrian Eon, is much longer and extends from
when the planet formed, about 4,550 million years (abbreviated "Ma"
for *mega anna*) ago, to the appearance of fossils of hard-shelled animals such as lobsterlike trilobites and various kinds of mollusks about
550 Ma ago. This eon is composed of two "eras," the older Archean
Era (from the Greek *archaios*, ancient) that spans the time
from 4,550 to 2,500 Ma ago, and the younger Proterozoic Era ("the
era of earlier life," from the Greek *proteros*, earlier, and *zoe*, life) that
extends from 2,500 Ma ago to the close of the Precambrian.

The younger and shorter eon is the Phanerozoic ("the eon of visible

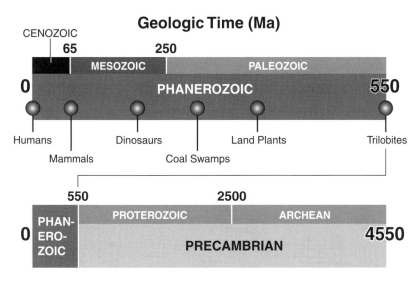

Figure 1.1 Principal divisions of geologic time.

life" from the Greek *phaneros*, visible or evident, and *zoe*). It encompasses the most recent Earth history, roughly 550 Ma, and is divided into three eras (from oldest to youngest: the Paleozoic, Mesozoic, and Cenozoic), each subdivided into shorter segments known as "geologic periods." The oldest such period of the Paleozoic Era (and, consequently, of the Phanerozoic Eon), spanning the time from about 550 to a little less than 500 Ma ago, is known as the Cambrian Period, named after Cambria, the Roman name for Wales, where rocks of this age (the Cambrian System of rocks) were first formally described. Until just a few decades ago, rocks underlying and thus older than those of the Cambrian System were universally regarded as lacking fossils. As far as anyone could tell, life of the Precambrian had left no trace.

During the early 1800s, rock strata of the Phanerozoic were first studied actively in northern Europe, mostly in Wales and England. For this reason, many of the Phanerozoic geologic periods and their systems of rocks are named after geographic areas or ancient peoples of what is now the United Kingdom. For instance the Ordovician and the Silurian, the two geologic periods sequentially younger than the Cambrian, are named in honor of the Ordovices and the Silures, two ancient Welsh tribes. The Devonian, the next youngest period of the Phanerozoic, is named for Devon, a county of southern England. And the Cretaceous, the geologic period famed for the extinction of the dinosaurs at its close 65 Ma ago, is named after the outcrops of dusty white-gray chalk (in Latin, *cretaceus*) that form the aptly named White Cliffs of Dover on the northern shore of the English Channel.

Younger Above/Older Below

The trailblazing geologists of the early 1800s who sought to subdivide Phanerozoic rock strata into a manageable series of geologic systems faced an immense challenge. They had none of the deep drill cores of today nor even easy access to many areas of well-exposed strata. Radioactivity, now the basis for the precise dating of ancient rocks, was unknown, not to be discovered until nearly a century later.

These geologic pioneers were forced to rely almost entirely on studies of beds that were exposed to view in naturally occurring rock outcroppings, exposures rare in the "green and verdant" British Isles, where bedrock is mostly hidden by plant cover. Soon, however— spurred by the Industrial Revolution—long, interlacing systems of canals, dug to connect port cities with centers of industry, brought to view extensive swaths of newly exposed rock strata. From this, one of the most straightforward and logical rules of geology soon became obvious: in any sequence of undeformed sedimentary rocks (those made up of sedimented debris, such as marine sandstones and silt-stones), any layer higher in the sequence must be younger than— deposited after—those below it. Known as the Law of Superposition, this simple notion—younger above/older below—provides a power-ful rule of thumb for determining the relative ages of geologic units.

Though it is a simple logical notion, the Law of Superposition has limitations. About 70% of the world's surface is covered by oceans, a setting where sands and silts slowly settle to produce the rock-form-ing sediment that coats the ocean floor. But these sands and silts are derived from the land masses, carried to the oceans by streams and rivers. This means that the rocks that make up the continents, the remaining 30% of the Earth's surface, are being weathered away and destroyed. Except in rare settings (for example, at the bottom of large inland lakes or in glaciated areas where rocky debris can be piled high by massive ice flows), the continental surface is a site of rock erosion, not of rock formation.

Rock weathering, coupled with the sporadic nature of sediment deposition, leaves a record full of gaps. Almost always, only a small fraction of the total time spanned by any given sequence of rocks is represented physically by the rocks that are actually preserved. Nowhere on the Earth is there a continuous rock sequence that preserves strata from all ages. Younger above/older below holds true (except where mountain building has turned rock units upside down), but because of the time gaps, exactly how much younger and how much older is hard to know.

The Globe as a Gigantic Jigsaw Puzzle

Gaps in the rock record are produced in other ways as well. Even the rocks that floor the world's oceans are not immortal. The main cause is "plate tectonics," the name applied to the movement of continental masses across the global surface, a process that leads inexorably to the destruction of ocean basin sediments.

The Earth's continents are like slowly moving pieces of a gigantic jigsaw puzzle. Consider in your mind's eye the outlines of the western coast of Africa and the eastern coast of South America. The two coastlines fit together (a match said to have been first noted in 1620 by English philosopher Francis Bacon): before the Atlantic Ocean was born, the two continents were actually joined as parts of a single, much larger supercontinental landmass called Gondwanaland. In the same way that a simmering pot of thick soup slowly churs as steam bubbles from its surface, the splitting apart and movement of such massive geologic plates is caused by heat escaping from the planet's interior through cracks in the Earth's crust. Because the rock masses are so immense, they move very slowly, on average only about 3 centimeters per year. But their movement is not smooth. At their boundaries they often stick together for a time and then suddenly jerk apart in the earth-rupturing jolts we know as earthquakes.

As the rocks of continents and ocean basins collide, the jerkily surging continental masses ride up over the sediments of the ocean floor, eventually forcing the sediments to such depths that they are melted by the Earth's heat. In this molasseslike state they ooze back

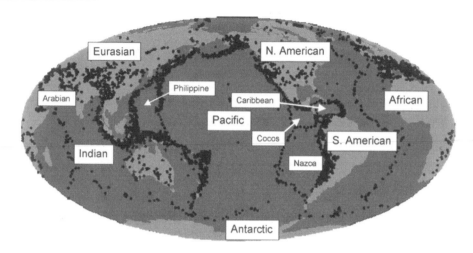

Figure 1.2 Like the shell of a giant egg, the Earth's surface is cracked into seven huge (and numerous smaller) interlocking tectonic plates outlined by strings of volcanoes (shown here by black dots).

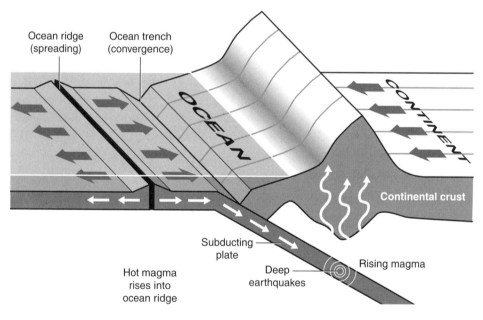

Figure 1.3 Where close-packed tectonic plates move away from each other, as at ocean ridges, hot magma wells up to congeal into strings of volcanoes that, if submarine, sometimes rise above the ocean surface, as in Iceland. Where plates collide, as at ocean trenches, the ocean floor is forced ("subducted") to depths so hot that it melts and then bubbles upward to form volcanic mountain chains such as the Andes and the Himalayas.

up through cracks and fissures in the crust and spew out at the surface as fiery lavas, giving rise to volcanic islands and ocean-rimming mountain chains like those that ring the Pacific Ocean from Fujiyama to the Aleutians and down the west coast of the Americas to the Andes. Large-scale continental movement, driven since the Earth's formation by heat escaping from its depths, has happened throughout geologic time. To us this movement is imperceptibly slow, but geologic time is so unimaginably long that no ancient rocks have survived on the ocean floor. The oldest known deposits of the world's ocean basins are geologically young, barely 250 Ma in age.

So, ordinary geologic processes—erosion, nondeposition, and the movement of the massive plates that make up the Earth's crust— conspire to limit the usefulness of the Law of Superposition. Without question, younger above/older below works well in a local outcrop or within a limited geographic area. But because this notion provides no way to gauge the length of time gaps in the preserved rock record, it cannot tell how much younger or how much older the rocks may be.

Gap-Filling Fossils

The geologists of the early 1800s who confronted these problems had to find some way to fill the gaps. They settled on a strategy of "stacking" the short sequences of rocks known from local areas into a single long column that would represent the entire rock record. Fossils were the key.

As more and more rock strata were studied, it became clear that different groupings of fossil species are present in rocks of different ages. The fossils could be used to order the rocks from older to younger. For instance, though trilobites and dinosaurs are not present together in any single rock unit, if both occur in a sequence of rocks the trilobites are always in strata below those that entomb the dinosaurs. The trilobites are demonstrably older, the dinosaurs assuredly younger. And as the science progressed, many different species, both of trilobites and dinosaurs were discovered—some more ancient, others more recent—and among these, certain forms were always present within, and therefore diagnostic of, rock strata of a particular previously well-established age. Wherever species of this type (termed "index fossils") are encountered, they provide firm evidence of the age of the embedding rocks.

Using the Law of Superposition and the insights provided by the documented succession of fossils and fossil assemblages, many local geologic sequences were soon linked together to make up a composite geologic column that by the 1830s already revealed the basic outlines of Phanerozoic evolutionary history. Enough detail was known to show even the two best-known episodes of Phanerozoic extinction—the demise first of the trilobites, at the end of the Permian Period and the Paleozoic Era, and later of dinosaurs at the end of the Cretaceous and the Mesozoic.

Studies of this type continue into the present, and a composite geologic column, based on countless geologic sequences, has been established for the entire Phanerozoic worldwide. This triumph is the result of two full centuries of observation, logic, and sheer hard work.

The "Schoolbook" History of Life

When we think of evolution, we think of the Phanerozoic history of life—the familiar progression from spore-producing to seed-producing to flowering plants, from animals without backbones to fish, land-dwelling vertebrates, then birds and mammals. Yet Phanerozoic rocks are like the tip of an enormous iceberg for they record only a brief late chapter—the most recent one-eighth—of a very much longer evolutionary story.

To see this, imagine that all 4,550 Ma of geologic time were condensed into a single 24-hour day. Evidence from the Moon and Mars tells us that for the first few hours of this "day" of Earth's existence its surface would have been uninhabitable, blasted by an intermittent stream of huge, ocean-vaporizing meteorites. At about 4:00 A.M., life finally gained a foothold. The oldest fossils were entombed in their rocky graves at 5:30 A.M. Gaseous oxygen—pumped into the environment by early-evolving plantlike microbes (cyanobacteria) and then chemically joined with oceanic iron to form rusty sediments known as banded iron formations—accumulated slowly until about 2:00 in the afternoon. Simple, floating single-celled algae with cell nuclei and chromosomes soon appeared, but by 6:00 P.M. they were supplanted

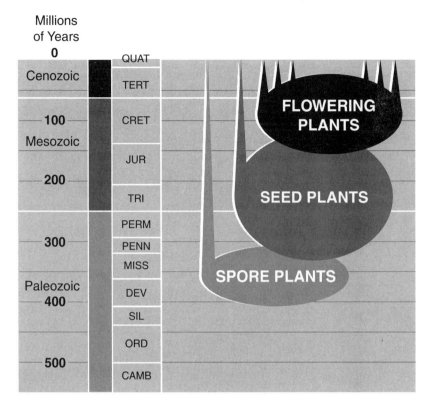

Figure 1.4 The Phanerozoic history of plant life.

by more rapidly evolving sexual plankton. At about 8:30 in the evening, larger many-celled seaweeds entered the scene and a few minutes later so did early-evolving jellyfish and worms.

The Precambrian, the period from the formation of the planet to the rise of shelled animals, spans 21 hours of this 24-hour "geologic day." The remaining 3 hours are left for the familiar Phanerozoic evolutionary progression, the schoolbook history of life recounted in texts and classrooms throughout the world. We humans arose only a few tens of seconds before midnight.

It is easy to understand why we might have a shortsighted view of life's long history and of the relevance of the Precambrian. The Phanerozoic fossil record has been a subject of fruitful study for more than two full centuries, and its organisms are large, striking, even

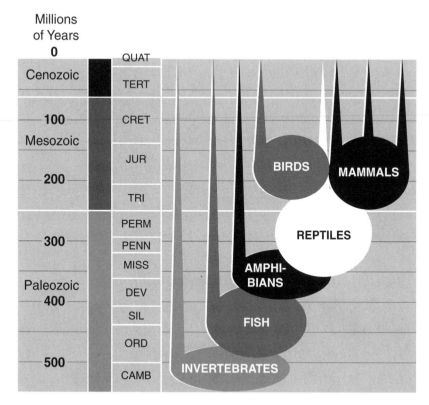

Figure 1.5 The Phanerozoic history of animal life.

awe inspiring. That this most recent 15% of Earth history is the "Age of Evident Life" is more than just a handy moniker. But studies of the Precambrian—the "Age of Microscopic Life"—have just begun. And though our relatedness to life of the Phanerozoic is obvious to all, it seems hardly credible that our roots extend to primitive Precambrian microbes, lowly life forms almost too small to be seen!

Yet obvious or not, each one of us is part of an evolutionary chain that extends to the distant Precambrian past and links us by the most basic living processes to ancient, primordial microbes. Why do we breathe oxygen? Why are we dependent on plants for the food we eat? Why is each of us similar to, but not identical with, our parents, our sisters, our brothers? The answers to these and other fundamental questions lie in an understanding of the Precambrian seven-eighths of

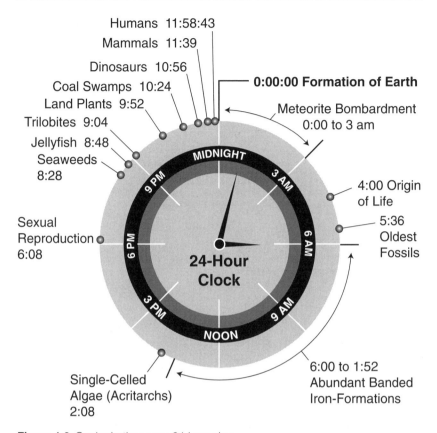

Figure 1.6 Geologic time as a 24-hour day.

the evolutionary story. And like so many aspects of natural science, the beginnings of that understanding come from the mid-1800s and the thoughts and writings of the famous British naturalist, Charles Robert Darwin (1809–1882).

Darwin's Dilemma

It should be obvious (though it often seems to be overlooked) that the human side of science—the personalities and status of particular scientists and the social setting in which they work—can and does have a long-term impact on the search for knowledge. The impact can be either positive or negative, pulling a field forward or tugging it back,

Figure 1.7 A portrait (by George Richmond) of Charles Robert Darwin when he was about 30 years old.

and at least in the short run its strength is often as dependent on a scientist's reputation as the factual arguments presented.

The history of the hunt for Precambrian fossils is a prime example. In 1859, in his epochal volume *On the Origin of Species*, Darwin first focused attention on the missing early fossil record and the problem it posed to his theory of evolution. As others took up the question—some in support of Darwin's views, others seeking to undermine them—the debate became contentious. Honest mistakes, unwarranted claims, promising finds, important discoveries were all made. But since there were few facts to go on, status and privilege played major roles in deciding whose view would win the day.

Of his newly minted theory, Darwin wrote:

> There is another . . . difficulty, which is much more serious. I allude to the manner in which species belonging to several of the main divisions of the animal kingdom suddenly appear in the lowest known [Cambrian-age] fossiliferous rocks . . . If the theory [of evolution] be true, it is indisputable that before the lowest Cambrian stratum was deposited, long periods elapsed . . . and that during these vast periods, the world swarmed with living creatures . . . . [However], to the question why we do not find rich fossiliferous deposits belonging to these assumed earliest periods prior to the Cambrian system, I can give no satisfactory answer. The case at present must remain inexplicable; and may be truly urged as a valid argument against the views here entertained.

Darwin's dilemma begged for solution. And though this classic problem was to remain unsolved—the case "seemingly inexplicable"—for more than 100 years, the intervening century was not without bold pronouncements, dashed dreams, and more than a little acid acrimony.

J. W. Dawson and the "Dawn Animal of Canada"

Among the first to take up the challenge of Darwin's new theory and its big problem, the missing early fossil record, was John William Dawson (1820–1899), a giant in the history of North American geology. For decades, Dawson was the most famous citizen of Montreal, Canada, and probably the only internationally recognized scientist in all of Quebec. Though not the official founder of McGill Uni-

Figure 1.8 Darwin as he is usually pictured—an elderly, heavily bearded, serious (or perhaps worried) scholar.

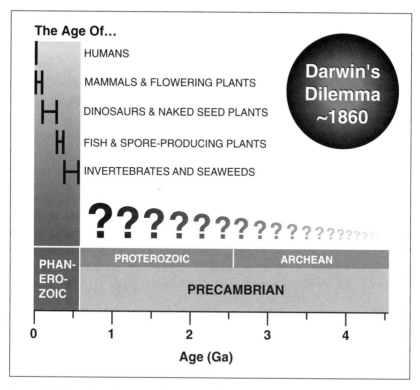

The Age Of...

- HUMANS
- MAMMALS & FLOWERING PLANTS
- DINOSAURS & NAKED SEED PLANTS
- FISH & SPORE-PRODUCING PLANTS
- INVERTEBRATES AND SEAWEEDS

Darwin's Dilemma ~1860

?????????????????????

PHAN-ERO-ZOIC	PROTEROZOIC	ARCHEAN
	PRECAMBRIAN	

0   1   2   3   4

Age (Ga)

Figure 1.9 Darwin's dilemma: the missing Precambrian fossil record.

versity, he was its first truly distinguished principal, appointed in 1855, and was largely responsible for its rise to prominence in Canadian education.

Dawson's kudos were many. He was the first president of the Royal Society of Canada (1882), president of the Geological Society of America (1895), and the only person to be elected president both of the American (1882) and British (1887) Associations for the Advancement of Science. In 1884, in his sixty-fourth year, he was knighted by Queen Victoria.

Sir John Dawson was well schooled, chiefly in Edinburgh, Scotland, where early in his career he became a protégé of Sir Charles Lyell, the most distinguished British geologist of the day (and a mentor and major influence also of Charles Darwin). Unlike Lyell, however, Dawson was an old-school Calvinist Christian and staunch anti-

Figure 1.10 The official portrait of Sir John William Dawson upon his appointment as principal of McGill University, Canada.

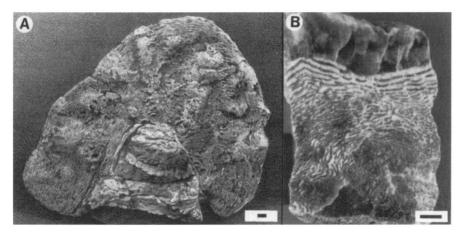

Figure 1.11 *Eozoön Canadense*, the "dawn animal of Canada," (**A**) as illustrated in Dawson's *The Dawn of Life* (1875), and (**B**) shown by a piece of the officially named (holotype) specimen archived in the National Museum in Washington, D.C. (Bars for scale represent 1 cm.)

evolutionist who, following centuries-old European tradition, believed that the proper, indeed the *only* role of science was to discover and exemplify, and thereby glorify, the workings of God. To J. W. Dawson, the obedient son of devout Scottish Presbyterians, science was a religious quest.

In 1858, a year before publication of *The Origin*, a local collector brought a number of unusual rock specimens to Sir William E. Logan, director of the Geological Survey of Canada. The specimens were old—gathered in Precambrian, "Laurentian" limestones (now dated at about 1,100 Ma) exposed along the Ottawa River to the west of Montreal—and they were evidently unique: thinly and regularly green- and white-layered in a way quite different from rocks typical of the region. Though Logan was not a paleontologist, the alternating regular layers suggested to him that these structures might be fossils. Over the next half-dozen years, Logan himself collected more specimens from limestone beds near Ottawa and displayed them at various scientific conferences, most notably at meetings of the Geological Society of London. Spirited debate ensued, but few geologists were willing to accept these curiously layered rocks as remnants of life.

In 1864, however, Logan brought specimens to Dawson, who not only confirmed their biologic origin but identified them as fossilized shells of huge foraminiferal protozoans ("forams" for short)—giant, oversized versions of the tiny, many-chambered shells that form thick accumulations in modern oceans and, when turned to rock, make up massive limestones like those used to build the great pyramids of Egypt. Dawson soon became so convinced of their biologic origin that a year later, in 1865, he proposed a formal scientific name for this supposed giant foram, *Eozoön Canadense*, the "dawn animal of Canada."

To many, Dawson's interpretation seemed hardly credible. Not only were these objects hundreds of times larger than any foram known, they were also evidently vastly older than any other fossils and were said to be preserved in rocks subjected to intense heating and deformation (metamorphism), sites quite unlikely to harbor fossil shells. But Dawson's published description of *Eozoön* was convincing, and the foraminiferal identity he proposed was soon seconded by William B. Carpenter, the leading foram specialist of the day. And since Carpenter was a senior scientist at the British Museum [Natural History], his pronouncement from this center of scientific power carried special weight.

With his view thus certified, Dawson became the champion Eozoonist, principal spokesman on behalf of what he termed this "remarkable fossil . . . one of the brightest gems in the scientific crown of . . . Canada," a role he continued to play up to his death some 40 years later. Over the intervening decades, claims and counterclaims, charges and countercharges—many intemperate, some highly personal—appeared in the public literature, most prominently as letters to the editor of *Nature* (today arguably the foremost science journal in the world but at that time a fledgling British periodical).

The first shot of the anti-Eozoonists was fired in 1866 by two Irish geologists—William King, an expert on Permian fossil animals, and Thomas H. Rowney, a noted mineralogist—who argued that they could find none of the traces of organic structure Dawson and Carpenter professed to see, and therefore regarded *Eozoön* as a purely mineralic, nonbiologic structure. Led by Dawson, the Eozoonist camp vehemently rebutted: King and Rowney hadn't seen the "best" specimens;

they obviously had written in haste; and, in any case, as neither foram specialists nor biologists they lacked the expertise to back their claim.

As the debate wore on, others joined the fray. In 1879, Karl Möbius, professor of zoology at the University of Kiel and the foremost foram expert in Germany, published a lengthy article in *Nature* that to many seemed to doom the Dawson-Carpenter claim. After studying some 90 sliced specimens of *Eozoön*, Möbius, like King and Rowney, was unable to detect telltale features of forams. He, too, was firmly convinced of its nonbiologic origin. By this time, Carpenter had withdrawn from the brawl, and though the anti-Eozoonists had gained the upper hand, Dawson was not to be dissuaded. He responded vigorously and at length (albeit with disdain and a degree of obfuscation).

Finally, in 1894, three full decades after Dawson had first seen specimens of this contentious "fossil," its claimed biologic origin was put to rest. J. W. Gregory and Hugh Johnston-Lavis discovered *Eozoön* in ejected blocks of limestone near Mt. Vesuvius in Italy and showed they were geologically young, formed by heat and pressure on limestone. No one, not even Dawson's steadfast supporters, could accept the notion of a "Precambrian fossil" shot out of a modern volcano!

The doubters had been right from the beginning. Dawson's famous and now infamous "dawn animal" was nothing more than a curiously layered mineral deposit formed when hot molten rocks intruded into Laurentian limestones, deforming and altering them to produce thin intermittent layers of a green-colored metamorphic mineral known as serpentine. Yet despite the overwhelming evidence, Dawson continued to press his case for the rest of his life. Over four decades, as a steady stream of evidence had mounted against him, he never admitted defeat.

Why had Dawson struggled on? In part because of hubris—dogged self-assurance. By all accounts, Dawson was a prideful man and, as the acknowledged expert on Canadian geology, his reputation was at stake. Moreover, he was pugnacious. The *Eozoön* controversy was just one of a number of contentious scientific issues in which he was a vocal participant. But Dawson's main spur appears to have been his strict Calvinist faith and unquestioning belief in biblical truth. Were

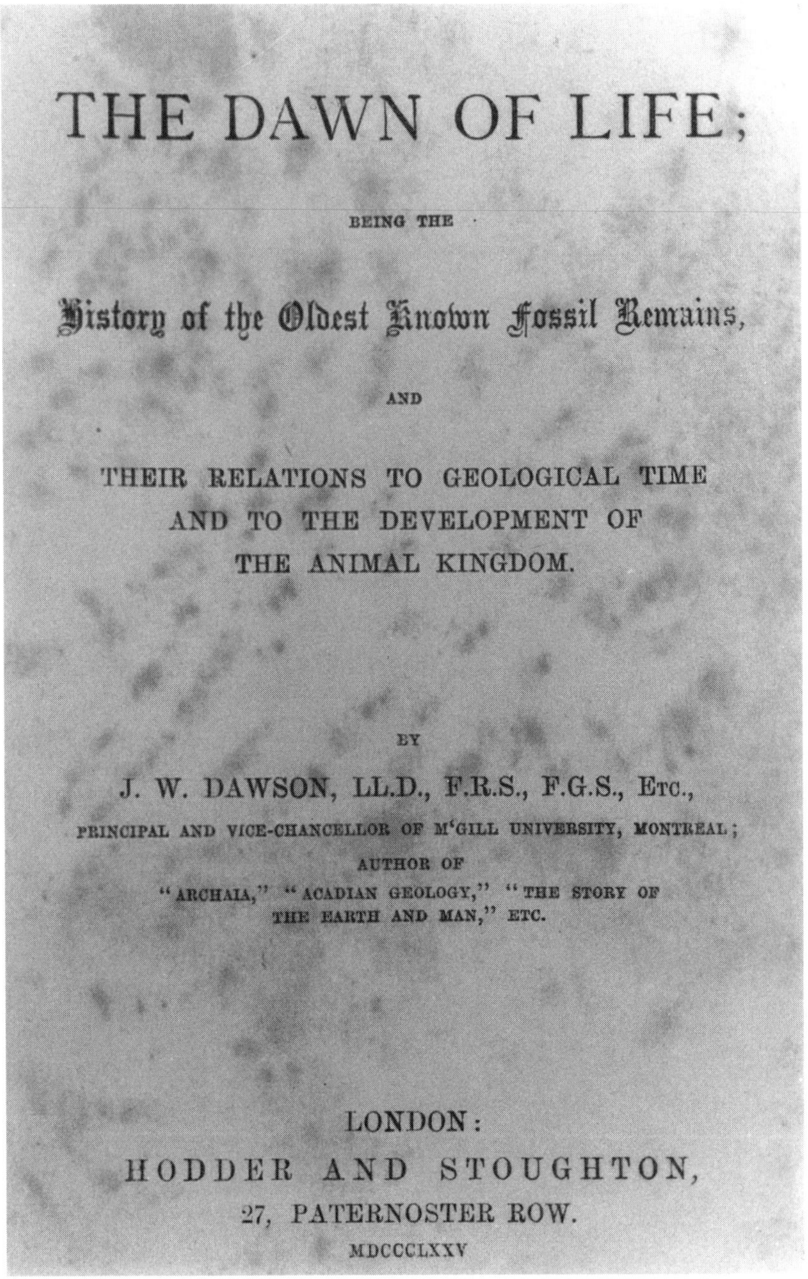

THE DAWN OF LIFE;

BEING THE

History of the Oldest Known Fossil Remains,

AND

THEIR RELATIONS TO GEOLOGICAL TIME AND TO THE DEVELOPMENT OF THE ANIMAL KINGDOM.

BY

J. W. DAWSON, LL.D., F.R.S., F.G.S., Etc.,

PRINCIPAL AND VICE-CHANCELLOR OF M'GILL UNIVERSITY, MONTREAL;

AUTHOR OF

"ARCHAIA," "ACADIAN GEOLOGY," "THE STORY OF THE EARTH AND MAN," ETC.

LONDON:

HODDER AND STOUGHTON,

27, PATERNOSTER ROW.

MDCCCLXXV

Figure 1.12 Title page of the first edition of J. W. Dawson's *The Dawn of Life*, published in 1875.

he to have proved the biologic origin of *Eozoön* he would have succeeded in exposing the greatest missing link in the entire fossil record, opening a gap so enormous he thought it would certainly undermine Darwin's theory by showing that evolution's claimed continuity was a myth, leaving biblical creation as the only answer. Dawson's view is well expressed in his book *The Dawn of Life*, written in 1875 when the debate was hot and heavy: "There is no link whatever in geological fact to connect *Eozoön* with the Mollusks, Radiates, or Crustaceans of the succeeding [rock record] . . . these stand before us as distinct creations. [A] gap . . . yawns in our imperfect geological record. Of actual facts [with which to fill this gap], therefore, we have none; and those evolutionists who have regarded the dawn-animal as an evidence in their favour, have been obliged to have recourse to supposition and assumption."

It is interesting—and ironic—that in the fourth and all later editions of *The Origin* Darwin cited the Precambrian age and primitive protozoal relations of *Eozoön* as consistent with his theory of evolution. This was just the sort of "supposition and assumption" that Dawson found so distressing.

Darwin was right in raising the question of the missing Precambrian fossil record. But he and Dawson were both mistaken in thinking that the "dawn animal of Canada" had bearing on the issue.

C. D. Walcott: Founder of Precambrian Paleobiology

Darwin posed the problem and others, most prominently J. W. Dawson, took up the challenge. But the first to make a lasting contribution to unraveling the mystery of the missing early fossil record was the American paleontologist Charles Doolittle Walcott (1850–1927), an impressively imposing figure: physically, at 6 feet 2 inches tall and an athletic build; intellectually, as discoverer both of the first Precambrian fossil cells known to science and the first excellently preserved Cambrian-age fauna; and administratively, as the most powerful science chief executive in America.

Like Dawson before him, Walcott was enormously influential and highly honored. Except for repeated forays into the field to do geology, he spent most of his adult life in Washington, D.C., where he served as the CEO of three of the most prominent of all scientific

Figure 1.13 Charles Doolittle Walcott, probably in 1894 at the time of his appointment as director of the U.S. Geological Survey. (Courtesy of E. L. Yochelson, Smithsonian Institution.)

organizations—first, as director of the United States Geological Survey (1894–1907), then as secretary of the Smithsonian Institution (1907–1927) and president of the National Academy of Sciences (1917–1923). Though his administrative responsibilities were heavy, he also found time to fit in terms as president both of the American Philosophical Society and, like Dawson, the American Association for the Advancement of Science.

Walcott was remarkably well connected. Appointed director of the U.S. Geological Survey by no less than President Grover Cleveland, he was well acquainted with Presidents McKinley, Taft, Harding, and Coolidge, a confidant of Presidents Theodore Roosevelt and Woodrow Wilson, and a close friend of captains of industry Andrew Carnegie and John D. Rockefeller. These contacts paid dividends to Walcott personally and to the organizations he headed as well as the country as a whole, for it was C. D. Walcott who influenced McKinley to set aside national forest reserves and who laid the groundwork with President Roosevelt for establishment of the U.S. National Park Service.

Surprisingly, however, Walcott had little formal education. As a youth in northern New York State he received but 10 years of schooling, first in public schools and, later, at Utica Academy (from which he did not graduate). He never attended college and had no formally earned advanced degrees (a deficiency for which he more than compensated in later life when he was awarded honorary doctorates by a dozen academic institutions).

In 1876, as a 26-year-old budding geologist and avid fossil collector, Walcott was hired as assistant to James Hall, chief geologist of the state of New York. This was a stupendous opportunity—Hall was the acknowledged dean of American paleontology, famous also in geologic lore as an "irascible tyrant" of "unbridled temper." Two years later, Walcott took a 2-week vacation to the town of Saratoga in eastern New York State to examine Cambrian limestone beds packed with distinctively layered, mound-shaped structures that Hall had discovered there a few years earlier. The more or less cabbagelike structures were just as Hall had described—irregularly shaped, large cannonball-like masses, up to a meter across and circular to oblong in cross section, made up of thin, undulating layers of dark- and light-

Figure 1.14 *Cryptozoon* reefs near Saratoga, New York. (Photo by E. S. Barghoorn, November 1964.)

colored limestone. Ever the astute observer, Hall believed the structures to have a decidedly "biologic look," and though he was never able to find direct evidence (such as preserved cells) of the organisms that constructed them, in 1883 he formally named them *Cryptozoon* (from the Greek meaning "hidden life") and interpreted them as reefs laid down by flourishing communities of microscopic algae. After a firsthand look at the specimens, Walcott too was convinced of their algal origin, an opinion that in later years was the foundation for his side of what came to be a rather nasty argument known as the "*Cryptozoon* controversy" (which we'll encounter again later in this chapter).

In the spring of 1879, the U.S. Geological Survey was established by an act of Congress. In July of that year, at the recommendation of James Hall, Walcott was appointed (officially, USGS employee no. 20) as an assistant geologist, the lowest rung on the Survey's ladder, receiving an annual salary of six hundred dollars. Immediately upon taking the oath of office he was assigned by survey director John Wesley Powell to an expedition heading to the vast, geologically little known Grand Canyon region of the western United States. Over the next several field seasons, Walcott and his comrades—geologists, naturalists, explorers, mappers, hunters—charted the geology of sizable segments of Arizona, Utah, and Nevada.

Director Powell, who had earlier led the first perilous expeditions into the depths of the Grand Canyon, had a hunch that Precambrian rocks there might contain evidence of ancient life. He assigned Walcott to assess the possibility. Powell's hunch bore fruit when, in 1883, Walcott reported he had found Precambrian specimens similar to *Cryptozoon*. This was a promising beginning, and over the next several years Walcott became convinced that his search for Precambrian fossils would eventually prove even more fruitful. In his words, from an article of 1891: "That the life in the [Precambrian] seas was large and varied there can be little, if any, doubt. . . . It is only a question of search and favorable conditions to discover it."

By 1899 Walcott's perseverance paid off with a startling find— fossils he thought were remnants of early animals: small, millimeter-sized black coaly disks found in Precambrian carbon-rich shales on the slopes of a prominent butte deep within the Grand Canyon. The

Figure 1.15 Walcott in the 1890s doing geological fieldwork from horseback in eastern California. (Courtesy of the U.S. Geological Survey and E. L. Yochelson, Smithsonian Institution.)

shales belonged to what is known as the Chuar Group of strata so, Walcott named the fossils *Chuaria*. The specimens were flattened, compressed between thin shale beds, and because Walcott (like Darwin before him) assumed that Precambrian rocks would yield the same sorts of fossils found in strata of the overlying Phanerozoic, he interpreted the disks as "the remains of . . . compressed conical shell[s]," possibly of marine invertebrate animals known as lampshells (brachiopods). In this he was mistaken. Rather than a small flattened shell, *Chuaria* is now known to be an unusually large, originally spheroidal, single-celled planktonic alga (technically, a "megasphaeromorph acritarch"). Nevertheless, Walcott's specimens were indeed authentic fossils, the first true cellularly preserved Precambrian organisms ever recorded.

After the turn of the century, Walcott moved his fieldwork northward along the spine of the Rocky Mountains, focusing first in the Lewis Range of northwestern Montana (an area since set aside as Glacier National Park) and then high in the Canadian Rockies at the eastern border of British Columbia in what is now Yoho National Park. From the Precambrian of Montana he reported many types of *Cryptozoon*-like structures (technically, "stromatolites"), all interpreted as built by "algae" (microorganisms today classed as cyanobacteria), to which he gave formal scientific names. And, in 1914 and again in 1915, he reported finding in these same rocks minute cells and chains of cell-like bodies he identified as fossil bacteria.

His studies in the Canadian Rockies, from 1907 to 1925, were even more rewarding. In 1909, near Burgess Pass at an elevation of 8,000 feet and close to the present-day tourist centers of Banff and Lake Louise, Walcott discovered a diverse marine flora and fauna that were amazingly well preserved in strata of Cambrian age, which he named the Burgess Shale. Between 1910 and 1917, when he was well into his sixties, he set up a quarry at the site and extracted literally tons of fossiliferous rock that he shipped to his Smithsonian laboratory for study.

Over the years, the remarkable fossils of the Burgess Shale have come to be increasingly famous, known to interested scientists and nonspecialists alike, as chronicled, for example, in Stephen Jay Gould's 1989 best-seller *Wonderful Life* and featured even on a December 1995 cover of *Time* magazine. Though many of Walcott's initial interpretations were later revised as other workers discovered new, astonishingly strange Burgess organisms (one especially bizarre and aptly dubbed *Hallucigenia*), his benchmark find has continued to provide the finest and most complete sample of Cambrian life known to science.

Walcott's contributions are legendary. He was the first discoverer in Precambrian rocks of *Cryptozoon* and other cyanobacterium-built stromatolites, of cellularly preserved algal plankton (*Chuaria*), and of possible fossil bacteria, all capped by his pioneering investigations of the world-famous fossils of the Burgess Shale! The acknowledged founder of Precambrian paleobiology (as the science has since come to be known), Walcott was the first to show, nearly a century ago and contrary to accepted wisdom, that a substantial fossil record of Precambrian life actually exists.

Seward's Folly and the *Cryptozoon* Controversy

The rising tide in the development of the field brought on by Walcott's discoveries was not yet ready to give way to a flood. Precambrian fossils continued to be regarded with skepticism, in part and perhaps mainly due to the aftertaste of Dawson's *Eozoön* debacle.

Figure 1.16 Evolution's "Big Bang" on the cover of *Time* magazine in December 1995.

Those reporting such finds were castigated as overly eager, biologically naive, geologically ignorant, or worse. A rather typical view was that of Harvard geology professor Percy E. Raymond, who in an acidic critique in 1935 characterized workers in the field as "so anxious . . . to obtain fossil from these [Precambrian] rocks . . . that any-

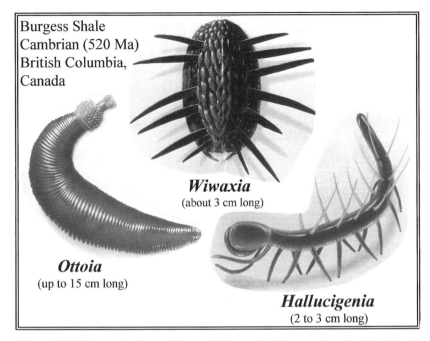

Burgess Shale
Cambrian (520 Ma)
British Columbia,
Canada

Wiwaxia
(about 3 cm long)

Ottoia
(up to 15 cm long)

Hallucigenia
(2 to 3 cm long)

Figure 1.17 Reconstructions of three of the nearly fifty species of animal fossils known from Walcott's Burgess Shale. *Ottoia*, a carnivorous worm and the largest of the three, ambushed prey from its burrow; *Wiwaxia* was a sluglike animal having scaly armor and bladelike spines that protected it from attack; *Hallucigenia*, a velvet worm, scurried about on spikelike legs, its back protected by needle-sharp spines.

thing which remotely resembles an organism is carefully saved and studied in the greatest detail." Since Raymond's sarcasm was directed primarily at Walcott (who had died a few years earlier), and since the two had a long history of antagonism, Raymond's analysis might be regarded as suspect. But he was not alone. Even harsher words were voiced by others, most prominently by A. C. Seward.

Charles Doolittle Walcott had brought Precambrian paleobiology to the brink of success. In one deft stroke, in a single influential textbook of 1931, Albert Charles Seward (1863–1941) snatched defeat from the jaws of this impending victory. "Seward's Folly" (that of A. C., not William Henry who bought Alaska for the United States from Russia in 1867) set the field into a tailspin of confusion that lasted for more than three decades.

Figure 1.18 Sir Albert Charles Seward on the occasion of his appointment as vice-chancellor of the University of Cambridge in 1924. (Courtesy of Cedric Shute, Natural History Museum, London.)

A. C. Seward was a man of intelligence and influence, many accomplishments, and a large number of titles. Associated throughout his career with the University of Cambridge, England, he served as professor of botany; master of Downing College; honorary fellow of Emmanuel, St. John's, and Downing Colleges; and, in 1924–25, as vice-chancellor of the university. He was a fellow both of the Linnean

Society (where Darwin's theory of evolution was first presented in 1858) and the Geological Society of London (where the biologic origin of *Eozoön* was hotly debated in the 1860s), and, like Sir J. William Dawson, he served as president of the British Association for the Advancement of Science. In addition to an earned Ph.D. degree he was recipient repeatedly of honorary Sc.D., D.Sc., and LL.D. degrees. As a fellow (and, from 1934 to 1940, as vice-president) of the prestigious Royal Society, he appended "F.R.S." to his signature, and, like Dawson, he was knighted (in 1936) by the British monarch.

Sir A. C. Seward was the most widely known and influential paleobotanist of his generation. He authored nearly a dozen paleobotanical textbooks, and because practially all claimed Precambrian fossils fell within the purview of paleobotany—whether presumed to be algal, like *Cryptozoon*, or even bacterial—Seward's opinion had enormous impact.

In 1931, in *Plant Life through the Ages*, the standard paleobotanical text then used throughout the world, Seward assessed the "algal" (that is, cyanobacterial) origin of *Cryptozoon* as follows:

> The general belief among American geologists and several European authors in the organic origin of *Cryptozoon* is, I venture to think, not justified by the facts. . . . [Such forms] are precisely the same in their series of concentric shells as many concretions which are universally assigned to purely inorganic agencies. . . . It is clearly impossible to maintain that all such concentrically constructed bodies are even in part attributable to algal activity. . . . [Cyanobacteria] or similar primitive algae may have flourished in Pre-Cambrian seas and inland lakes; but to regard these hypothetical plants as the creators of reefs of *Cryptozoon* and allied structures is to make a demand upon imagination inconsistent with Wordsworth's definition of that quality as "reason in its most exalted mood."

Seward was even more categorical in his rejection of Walcott's report of fossil bacteria:

> In a very few examples [of *Cryptozoon*-like structures] the residue left after treating the rock with acid revealed the presence of a small number of cell-like structures, the organic nature of which cannot be said to

have been established. . . . It is claimed that sections of a Pre-Cambrian limestone from Montana show minute bodies similar in form and size to cells and cell-chains of existing [bacteria]. . . . These and similar contributions . . . are by no means convincing. . . . We can hardly expect to find in Pre-Cambrian rocks any actual proof of the existence of bacteria.

An entertaining writer whose texts are spiced with scholarly quotes and poetic couplets, Seward closed his discussion with the following:

My desire is to lay stress on the need of a more critical examination of the evidence which has led to the description of the earliest phase of geological history as an "Age of Algae"—algae with doubtful credentials:

"Creatures borrowed and again conveyed, From book to book—the shadows of a shade."

Seward was partly correct. Mistakes *had* been made. Mineralic, purely inorganic objects *had* been misinterpreted as fossil. Better and more evidence, carefully gathered and dispassionately considered, *was* much needed.

But, overall, Seward was in error. His aggressive skepticism, delivered from his throne of unquestioned authority, was a disservice to the field. Seward's positive contributions were many (for he truly was a leader in the development of paleobotanical science), but his assertive disbelief of the biologic origin of *Cryptozoon* and his dogmatic pronouncement that "we can hardly expect to find in Pre-Cambrian rocks any actual proof of the existence of bacteria" stifled the hunt for the missing Precambrian fossil record for nearly 40 years.

Denouement

Darwin, Dawson, Walcott, and Seward—these were the movers and shakers during the embryonic first century of search for the missing Precambrian record of life. From time to time they were abetted by others, but it was they who set the course of the fledgling field as it felt its way in fits and starts toward the present.

2

Birth of a New Field of Science

The Floodgates Crack Open

In the the mid-1960s—a full century after Darwin broached the problem of the missing early fossil record—the hunt for early life began to stir, and in the following two decades the floodgates would finally swing wide open. But this surge, too, had harbingers, now dating from the 1950s.

A Glimpse through the Precambrian Metamorphic Veil

In 1953, a prominent American economic geologist, Stanley A. Tyler (1906–1963) of the University of Wisconsin, set out to investigate the geology of the Gunflint Formation, a mid-Precambrian (2,100-Ma-old) iron-rich rock unit that straddles the U.S.-Canada border between northern Minnesota and southern Ontario. Funded by the newly formed U.S. National Science Foundation, his goal was to determine the geographic spread of the ore-bearing beds and how the iron had been laid down. He began his study at the then active ironworks of Minnesota's Mesabi Range near the mining town of Hibbing and headed northeast along the western shore of Lake Superior, tracing the strata to their farthest reach in Ontario, a distance of some 500 kilometers.

One Sunday in late August Tyler took a day off, rented a dinghy and outboard motor, and went fishing near Flint Island up the coast from the little lakeside village of Schreiber, Ontario. As he cast and reeled his line, he spotted an odd-looking outcrop of Gunflint rocks on the nearby shore. He pulled the boat onto the rocky shingle (a site later dubbed the "Schreiber Beach Locality") and took a look.

Figure 2.1 Stanley A. Tyler in 1959, teaching students about the Proterozoic Baraboo Quartzite at Van Hise Rock in the Baraboo Hills of Wisconsin. (Courtesy of John Valley and Robert Dott, University of Wisconsin, Madison.)

The flat shelflike outcrop, extending a few tens of meters back from the water's edge, was slightly inclined, sloping gently into the lake. Tyler recognized immediately that this was a bedding plane exposure—glaciation of the great Ice Age and countless Lake Superior

Figure 2.2 Location of the Schreiber Beach site of the Gunflint chert, on the northern shore of Lake Superior in southern Ontario, Canada.

winters had stripped away overlying strata, bringing to view the upper surface of a half-meter-thick bed of dense, fine-grained chert, a type of sedimentary rock composed of interlocking grains of the mineral quartz, SiO_2. But the rocks were unlike the rusty-red, iron-rich cherts typical of Gunflint beds elsewhere. These were jet black, and their distinctive waxy, glasslike luster suggested the quartz grains that made them up were extremely small. He was further surprised to see that the chert bed was packed with dozens of closely spaced *Cryptozoon*-like mounds, some more than a meter across, each built up of a nested series of more or less concentric thin wavy layers. Clearly, these iron-lacking black cherts were unusual. Though Tyler was chiefly interested in the iron-bearing beds of the formation, he collected several hand-sized specimens from the outcrop.

Whether Tyler landed any lake trout that afternoon is unknown, but the distinctive jet black cherts he bagged on this fateful outing would soon prove a remarkable catch.

When he returned to his laboratory at the end of the field season, Tyler selected a suite of rock samples to be prepared for further study.

Figure 2.3 Gently inclined bedding plane exposure of the Gunflint chert at Schreiber Beach. (Scale shown by the geologic hammer at left-center.)

Each specimen was sliced into millimeter-thin slabs, which were then cemented onto glass microscope slides and ground to a waferlike thinness—preparations, known as petrographic thin sections to be studied using a high-powered microscope. The iron-poor black cherts from Schreiber Beach were almost an afterthought, but as he had suspected at the outcrop he found them to be composed of exceedingly tiny quartz grains, the largest only a few micrometers (one thousandth of a millimeter, abbreviated "μm") in size. This meant that unlike the Gunflint strata to the far west, in northern Minnesota, those at Schreiber Beach had escaped the pressure-cooking of geologic metamorphism, in Tyler's phrase "providing a glimpse through the Precambrian metamorphic veil."

He solved easily the question of their black color, for in the thin sections he could see that the rocks contained abundant cloudlike clumps and wispy layers of microscopic particles of dark brown to black coaly organic matter. The rocks looked like a chertified very thinly layered deposit of black coal. But he was mystified to discover that each thin section was packed full of layer upon layer of tens of

Figure 2.4 Circular, *Cryptozoon*-like, microfossil-bearing Gunflint stromatolites at Schreiber Beach.

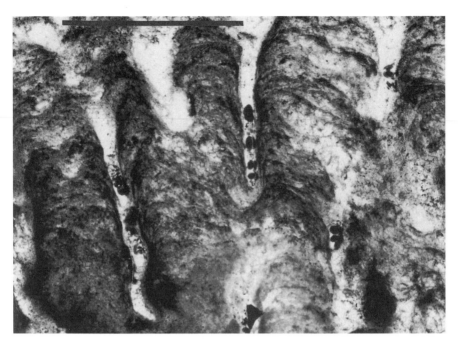

Figure 2.5 Vertically sliced thin section of the Gunflint chert showing the wispy, dark brown, organic-rich layers formed by stromatolite-building microorganisms. (Bar for scale, upper left, represents 1 cm.)

thousands of clearly defined, uncompressed, long, thin, threadlike filaments and tiny, hollow, balloonlike balls. Never before had he seen anything like this!

Tyler was an expert mineralogist. He was sure that the minute brownish threads and spheres were not mineral grains. And in the thin sections he could see that the tiny objects were totally embedded in the cherts, enclosed on all sides by quartz grains, so they certainly were not microscopic contaminants introduced in the laboratory. But if they were neither mineral grains nor laboratory contaminants, what were they? Their texture and dark brown color suggested that they were composed of coal-like organic matter. If so, the sinuous interwoven strands and small hollow balls were likely to be some sort of microscopic fossils. This was a total surprise.

Questions raced through Tyler's mind. The mid-Precambrian age of the Gunflint rocks seemed established beyond question. If these really

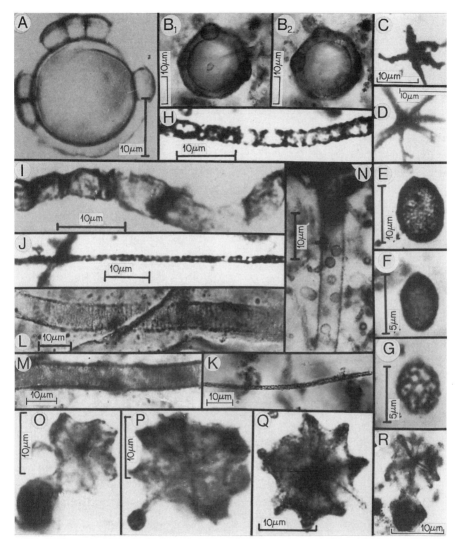

Figure 2.6 Microfossils of the Gunflint chert. (**A** and **B**) *Eosphaera*, in (B) shown in two views of the same specimen; (**C** and **D**) *Eoastrion*; (**E**, **F**, and **G**) *Huroniospora*; (**H** through **K**) *Gunflintia*; (**L** and **M**) *Animikiea*; (**N**) *Entosphaeroides*; (**O** through **R**) *Kakabekia*.

were microscopic fossils, he had made a stupendous discovery. But how could he be sure? Perhaps the "fossils" were merely soil microbes that had somehow worked their way into minute cracks and fissures in the cherty sediments. Or maybe they were a fluke, some type of purely mineralic oddity he simply didn't know about. And if they actually were fossils, what type of fossil organisms might they be? He checked the textbook from the one paleontology course he had taken during his student days, but he was unable to find anything that even remotely resembled the tiny threads and spheres.

Tyler vacillated between being pleased and puzzled. All of his experience and all of the evidence he could muster indicated that the distinctive microscopic objects were not mineral, that they must be ancient fossils. This was his best guess. But it was only a guess. He was a mineralogist—not a biologist, not a paleontologist—and he was determined not to climb out on this shaky limb without getting expert confirmation.

Early in the fall of 1953, only a few weeks later, Tyler took with him photographs of his microsopic finds when he journeyed to Boston to attend the national meeting of the Geological Society of America, a yearly gathering of several thousand professional geologists for the purpose of exchanging ideas and renewing old friendships. There, at the annual cocktail party of the faculty and geologic alumni of the Massachusetts Institute of Technology (MIT), he sought out Robert Shrock, co-author of the one paleontologic textbook Tyler owned, and showed him pictures of the Gunflint "fossils." Immediately, Shrock became excited. After only a few moments he not only was convinced that the objects were true microfossils but had identified them as parts of simple fungi—the slender filaments as the tubular vegetative bodies of the organisms and the small cell-like spheres as reproductive spores. As Shrock told me years later, they were "exactly the sort of fungi that cover the top of a jam jar left open too long."

Shrock was a highly distinguished paleontologist, known worldwide, but because he was an expert on fossil animals (especially invertebrates, those without backbones, like corals and clams) he felt it important to have his assessment firmed up by someone knowledgeable about ancient plants. As luck would have it, Shrock knew exactly the right person. Elso S. Barghoorn (1915–1984), a bright young paleobotanist, had just recently been appointed to the biology faculty at Harvard, a short sprint up Massachusetts Avenue from MIT. Not only did Barghoorn know about plant fossils and geology, prerequisites for

Figure 2.7 Elso S. Barghoorn in 1964 at a cottage near Schreiber, Ontario, the base camp for his geologic studies of the Gunflint chert.

interpreting Tyler's fossil-like objects, but he also was a fungal specialist, an expertise he had developed in Panama during the Second World War when he studied the microscopic filamentous fungi that were fouling binoculars and other military equipment in the Pacific theater.

With Tyler in tow, Shrock headed toward the Society's Harvard cocktail party to find Barghoorn. By chance, halfway there, they met Barghoorn in the lobby of the Statler Hotel. The three—Tyler, Shrock, and Barghoorn—sat, chatted, and plotted a research strategy.

By mid-January, a few months later, Tyler had sent Barghoorn several photographs and a one-paragraph summary of his research results. Barghoorn fleshed this out with a set of biologic interpretations, and their short manuscript reporting the discovery of the Gunflint fossils was published in the April 30, 1954, issue of *Science*, the journal of the American Association for the Advancement of Science.

This was a very sketchy report, assembled in great haste, and it is perhaps not surprising that though "five morphologically distinct" types of fossil organisms were said to have been identified (one kind of flagellated protozoan, two types of algae, and, following Shrock's lead, two types of fungi), only one of these—a form compared with "algae" (cyanobacteria) of the genera *Lyngbya* and *Oscillatoria*—has stood the test of time. And in order to avoid the still smoldering controversy about the origin of *Cryptozoon*, they elected not to mention that the densely packed fossils were found within and actually made up the concentrically stacked layers of *Cryptozoon*-like mounds (an association that, once recognized, would prove pivotal to the development of the field). Nevertheless, this article on "the oldest structurally preserved organisms that clearly exhibit cellular differentiation and original carbon complexes which have yet been discovered in pre-Cambrian sediments," was a benchmark, a monumental "first."

Unheralded Contributions of a Soviet Bear

At about the same time, in the mid-1950s, a series of articles by Boris Vasil'evich Timofeev (1916–1982) and his colleagues at the Institute of Precambrian Geochronology in Leningrad (now St. Petersburg) reported the discovery of microscopic fossil spores in Precambrian siltstones of the Ural Mountains, the Ukraine, and elsewhere in the Soviet Union. Though part of these reported finds would ultimately gain widespread acceptance, their initial reception in the West—for both scientific and political reasons—was as frigid as the then raging Cold War.

Timofeev was a burly bear of a man, but more of a disheveled teddy bear than the growling Soviet bear pictured in much of the noncommunist world. He was a gracious host, both in his laboratory and his small apartment, and in his office (with the door closed) he delighted in serving visitors tumblers of 200 proof lab alcohol laced with one or another of the flavored tinctures kept under lock and key in his desk drawer. A Ukrainian by birth and of friendly demeanor, he had risen by dint of effort—despite an earlier stay in one of the infamous gulags—to become head of the Precambrian Paleontology Laboratory of his institute, a position of considerable responsibility.

In thin sections, like those studied by Tyler and Barghoorn, fossils

Figure 2.8 Boris Vasil'evich Timofeev in 1969, during a geologic field trip to collect Precambrian siltstones in western Ukraine. (Courtesy of Mikhail A. Fedonkin, Paleontological Institute, Russian Academy of Sciences, Moscow.)

are detected *within* the rock, entombed in the mineral matrix, so the possibility of laboratory contamination can be ruled out (see plates 1 and 2). But preparation of thin sections requires special equipment, and their microscopic study is tedious and time consuming. A faster technique, pioneered for Precambrian studies in Timofeev's lab, is to concentrate the fossils by dissolving the rock in mineral acid (hydro-chloric acid for limestones, hydrofluoric for cherts and siltstones). Be-cause of their coaly composition, the organic-walled microfossils pass through the technique unscathed. Abundant fossils are concentrated in the resulting sludgelike acid-resistant residue (known technically as a palynological maceration) which can be slurried onto a microscope slide for study (see plates 3 and 4).

Unfortunately, the maceration technique (patterned after that em-ployed by Soviet petroleum geologists to extract fossil pollen and spores from Phanerozoic shaley rocks) is subject to error-causing con-tamination. Contaminants can be introduced at almost every stage of the process. At the beginning, even though rock surfaces are cleaned with care to remove sticky soil, microbes in tiny rock crevices are likely to be missed. Laboratory water and commercially available mineral acids can also contain a zoo of living contaminants—bacte-ria, cyanobacteria, unicellular algae, microscopic fungi. And an al-most limitless array of fossil-like objects can be introduced during transfer of the residue onto microscope slides. Common culprits in-clude dust, cigarette ash, spores, and pollen grains that settle from the lab air; lint fibers from clothing or the cloth used to clean microscope slides; small woody fragments and chunks of resin abraded off the wooden rods used to stir the acid-rock sludge; flakes of dandruff and strands of hair; even bits of small bugs (evidently, parts of spiders that live in water pipes).

From the 1950s into the 1970s, as studies of the Precambrian fos-sil record were getting under way in earnest, all of these maceration-borne contaminants were misinterpreted as fossils by one worker or another. But it is important to remember that during these years, the true nature of Precambrian life was utterly unknown. There was no established fossil record with which to compare new finds. No one knew what to expect. Mistakes were easy to make and Timofeev's laboratory was not immune. Moreover, though Timofeev had sound geologic training and a lot of field experience, his biologic back-ground was wanting. Yet he had no chance to fill in the gap—for years he was practically the only Soviet worker in the field, and be-cause of his Gulag past his correspondence with foreign scientists was closely monitored, contact with foreign visitors restricted, and travel outside the Soviet Union prohibited.

Viewed in this light, it is no wonder that much of Timofeev's early work was less than sterling. Some of the "fossils" he reported were contaminants. Others were Phanerozoic rather than Precambrian in age. Still others were algal phytoplankton, not spores of land plants as he had initially supposed. And because publishing photographs at that time in the USSR was nearly impossible for those who were not members of the scientific elite, practically all of Timofeev's early studies were illustrated by his own line drawings, a form of presentation notoriously subject to self-delusional "fact fudging."

The 1950s, the early days of Timofeev's work, spanned the height of the Cold War. In the United States, Joseph R. McCarthy, the frenetic junior senator from the state of Wisconsin, led the charge to cleanse the country of communists, pinkos, and so-called fellow-travelers. President Eisenhower and the American Congress set aside funds to build the interstate superhighway system, prodded by the notion that it would serve as an escape route from urban centers during nuclear attack. Schoolchildren were taught to "duck and cover." Television, a new addition to affluent homes, overflowed with warnings of the "Red Menace." In the West, things Soviet were viewed with distrust. Soviet science was no exception (to many a view well justified by the earlier Lysenko fiasco in Soviet genetics), and the antipathy could be deep seated. Barghoorn, for instance, refused ever to visit the Soviet Union and was loathe even to acknowledge Soviet scientific articles in his research reports. (This position understandably became rock solid in the early 1960s when his brother, Frederick, a Yale social science professor, was imprisoned in Moscow as an alleged spy, and finally freed—after two weeks of sleep-deprived interrogation—as a result of repeated personal pleas by President Kennedy to Chairman Khrushchev).

Ultimately, of course, firmly established facts win out in science. But in the short run, acceptance of a new idea can be influenced decisively by the prevailing political climate and the prestige of the scientist proposing it. Whether right or wrong, on the mark or not, the views of Darwin, Dawson, Walcott, and Seward all received respectful hearings because they were espoused by internationally recognized luminaries, and Barghoorn's position at Harvard added a telling air of authority to the Tyler-Barghoorn report of fossils in the Gunflint chert. In contrast, neither Timofeev, his work, his laboratory, nor his institute was well known beyond the boundaries of the Soviet Union, and acceptance of his views in the West was poisoned by international politics. Though some of Timofeev's work was flawed, a large part has proved sound and the technique he pioneered to discover microfossils in Precambrian shaley rocks is now used worldwide. Yet in the 1950s it probably would not have mattered had Timofeev discovered the Rosetta Stone of this or any other science. Beyond the confines of the Iron Curtain, some would have found reason to doubt his claims, no matter what!

Famous Figures Enter the Field

Early in the 1960s, the fledgling field was joined by two geologic heavyweights, an American, Preston Cloud, and an Australian, Martin Glaessner, both attracted by questions posed by the abrupt appearance and explosive evolution of shelly invertebrate animals that mark the start of the Phanerozoic Eon. Each had been interested for some years in this classic "Precambrian-Cambrian boundary problem," Cloud as early as 1948 and Glaessner at least since the mid-1950s.

The Wiry Wonder

Preston Cloud (1912–1991) was neither tall nor stout. In fact, at a height of about 5 feet 6 inches and a lean build of perhaps 135 pounds, to some he may have appeared diminutive. But he never was. In any way. Cloud was a giant, a wiry wonder, full of energy, ideas, opinions, and good hard work. And he was probably the greatest biogeosynthesist the United States ever produced.

Born in West Upton, Massachusetts, Cloud worked his way up from modest beginnings. Perhaps because of this he was feisty, a fighter figuratively and evidently literally as well. He is said to have been bantamweight boxing champion of the American Pacific Scouting Force during World War II, and though I cannot vouch for the claim, the image fits. He was a person who sought to dominate, a no-nonsense

Figure 2.9 Preston E. Cloud, about 1975, at a scientific meeting at Snowbird, Utah.

leader who did not suffer fools gladly. He once told me that when he was appointed chief of the Paleontology and Stratigraphy Branch of the U.S. Geological Survey in 1949, he had his chair and desk placed on 4-inch-high risers so he could look down on those coming to him to plead their cases—a position of authority he was convinced helped him do his job. His 10-year stint in this post was notably successful as he doubled the number of geologists in the branch and raised it to a level of scientific preeminence never seen before or since.

At work, Cloud was not given to idle chatter and struck some colleagues as a bit imperious (one of them referred to him as "the little general," though never to his face). Yet Cloud had an overriding saving grace. He was brilliant! His Precambrian interests were first evident in the late-1940s when he argued in print that though the known Early Cambrian fossil record was woefully incomplete, it was the court of last resort and, ultimately, the only court that mattered. Cloud's view was that any and all notions regarding the Cambrian explosion of many-celled animals should be based on painstaking examination of the fossil record as known, on directly available hard data rather than mushy wishful thinking.

In the 1960s, Cloud became much more active in the field, writing a major paper that to many certified the authenticity of the Tyler-Barghoorn Gunflint microfossils. Later he authored a series of reports adding new knowledge to the Precambrian record of microbial life. But Cloud's interests were broad and eclectic. He was a geologist to the core and knew that the rock record held the key to understanding the evolution not only of life but of the Earth's environment. And he was a gifted synthesist, showing his mettle in a masterful paper of 1972 ("A Working Model of the Primitive Earth"), where he set the stage for modern understanding of the interrelated atmospheric-geologic-biologic history of the Precambrian planet.

Preston Cloud was a true giant in the development of Precambrian paleobiology.

The Austrian Australian

Along with Tyler and Barghoorn, Timofeev and Cloud, there was one more prime player in this now fast-unfolding field, Martin Glaessner (1906–1989) of the University of Adelaide in South Australia. A scholarly, courtly, old-school professor of geology and the internationally acclaimed "father of modern micropaleontology" (so regarded since publication in 1945 of a classic textbook on the subject), Glaessner was the first to make major inroads toward understanding the (very latest) Precambrian record of many-celled animal life.

Born on Christmas day in 1906 in northwestern Bohemia, Glaessner was educated at the University of Vienna, where he earned doctorate degrees both in law and in science (like Charles Lyell, he supposed that income from the practice of law might be needed to support his early-found passion for natural history). By the time he received his second doctorate, at the age of 25, he had already published nearly a

Figure 2.10 Martin F. Glaessner, about 1988. (Courtesy of Brian McGowran, University of Adelaide, Australia.)

score of scientific articles and was invited to Moscow to organize research in micropaleontology for the Petroleum Research Institute of the Soviet Academy of Sciences. There, in 1933, Glaessner met the ballerina Tina Tupikina. They were wed 3 years later and, spurred by the Soviet regime's dictum that foreign specialists must either take up USSR citizenship or leave, he and his young wife departed Moscow for Vienna in 1937.

Only a few months later Hitler's army occupied Austria. With the help of friends in London, Glaessner and his wife were soon again on their way, this time to Port Moresby, New Guinea, where he was to organize a micropaleontological laboratory for the newly formed Australasian Petroleum Company. And when war came to New Guinea in 1942, the Glaessners fled once more, this time to Australia.

In 1946, Glaessner was awarded his third doctorate, honorary doctorate of science at the University of Melbourne, and in 1950 he joined the faculty of the University of Adelaide.

Three years before Glaessner was appointed professor at Adelaide, Reginald C. Sprigg announced his discovery of fossils of primitive soft-bodied animals, chiefly imprints of saucer-sized jellyfish, in the Ediacara Hills of South Australia. Though Sprigg first thought that the fossil-bearing beds were Cambrian in age, Glaessner showed them to be Precambrian (albeit marginally so), the oldest fossils of multi-celled animals known to science. Together with his colleague, Mary Wade, Glaessner spent much of the rest of his life working on this benchmark fauna, bringing it first to international attention in a *Scientific American* paper of the early 1960s, and later in a landmark monograph, *The Dawn of Animal Life*, that appeared in 1984.

With Glaessner in the fold, the stage was set. Like a small jazz band—Tyler and Barghoorn trumpeting microfossils in cherts, Timofeev beating on fossils in siltstones, Cloud strumming the early environment, Glaessner the earliest animals—great music was about to be played. At long last, the curtain was to rise on the missing record of Precambrian life!

A Youngster Joins the Fray

In the fall of 1960, when I first became fascinated with this area of science, I knew nothing of the foregoing. I was young, an 18-year-old sophomore at Oberlin College in northeastern Ohio. That semester I was enrolled in my second geology course, "History of the Earth," and I listened intently as my favorite professor, Larry DeMott, raised the question of the missing Precambrian fossil record and the problem it posed to Darwin's evolution.

For some reason, now lost to me, the matter struck me as extraordinarily intriguing. I had been brought up in a family of scientists (my mom was schooled in botany and mathematics, my dad was a paleobotanist), and there was no doubt in my mind that Darwin was right. Evolution was a fact, so there simply *had* to be a Precambrian fossil record. Its absence might have been "inexplicable" to Darwin, but I soon became determined that it would not be so to me.

Figure 2.11 A youthful Bill Schopf, in 1970, two years after completing his graduate studies, holding vials of Moon dust collected during NASA's Apollo Program.

At that time, Oberlin had probably the largest small-college library in the world. I read everything I could find about Precambrian life. The more I read the more enamored I became, and within a few months completed a forty-page essay on the subject (a sophomoric synthesis which, remarkably, turned out to be the outline for my life's work). I was particularly taken by a passage in a slim 1949 volume by the renowned evolutionary biologist George Gaylord Simpson. Simpson argued that because the evolutionary distance between humans and trilobites seemed roughly the same as between a trilobite and an amoeboid protozoan (then thought to be among the earliest forms of life), and because the oldest trilobites were about 500 Ma in age, then the first amoebas—and thus the origin of life itself—must date from about 1,000 Ma ago. This would mean that the origin of life had required an enormously long period—billions of years. But to Simpson this made good sense because he thought that the distance between non-life and the first organisms must be vastly greater than between any two kinds, fossil or modern.

Clearly, a lot of this was guesswork. But if Simpson's notion was even close to being right, it told me that all I had to do was to trace back the fossil record to about 1,000 Ma ago, where I might then expect to find direct evidence of life's beginnings. I was young, naive, and full of enthusiasm. This was wonderfully heady stuff. I was sold!

Well-Intended Words of Caution

The following April, while I was home from college during spring recess, my dad took a visiting young British paleobotanist, Bill Chaloner, on a field trip to southern Ohio to collect fossil plants from a Devonian-age black shale. My older brother, Tom, and I tagged along. As we sat on the outcrop, splitting rocks and searching for shiny black bits of fossils, the wind came up and it began to rain, first in dribs and drabs and then by the bucketful. My dad and Tommy huddled under one of our army surplus ponchos and Chaloner and I under the other. Searching for tiny black slivers of fossil plants on the surface of rain-soaked black shale slabs is about as promising as trying to catch a black cat in a cave with your eyes closed. We soon gave up the fossil hunt.

Isolated on that outcrop in the midst of a swirling downpour, we began to chat, Chaloner asking me about my plans. With heartfelt enthusiasm, I told him about my bold hopes to uncover the missing Precambrian record of life. Though only a recent Ph.D., Chaloner was experienced and worldly (and has since become Britain's most distinguished paleobotanist and a fellow of the Royal Society). And, like all British paleobotanists (indeed, like paleobotanists worldwide), Chaloner was thoroughly schooled in the writings of A. C. Seward. He began his reply gently, thoughtfully, telling me that this was "a good problem" but that he "wanted me to succeed," and since this question had gone unanswered for more than a century it would be "prudent" for me to approach it with "caution."

His kind advice was plain—after graduating from college I should earn a master's degree and a doctorate (working on some "potentially solvable scientific problem") and then find appointment as an assistant professor, which, if all went well, would lead to a tenured position on a university faculty. As a tenured professor, I would be free to work on anything that interested me, even something that stood such an astonishingly slim chance of success as finding Precambrian fossils.

I quickly tallied up the years—college (2 ¼ to go), graduate school (probably 6), and assistant professor (6 more). I was 18 years old, and Chaloner was proposing that I wait another 14 or more years—almost my whole lifespan!—then, "if all went well," I could finally embark on my quest. My dream was crumbling before my eyes!

The Fire Burns On

Despite this well-meant advice, I still had the Precambrian fire in my belly. The next fall I screwed up my courage and wrote to the only two Americans in the field I had managed to identify, Elso Barghoorn at Harvard, and Preston Cloud, the newly appointed head of the Department of Geology at the University of Minnesota. My hope was that one or the other would find me suitable as a prospective graduate student.

Both treated me with kindness, and Barghoorn even gave me chunks of the Gunflint chert which I sectioned and used as the subject of my Oberlin honors thesis. In the summer of 1963, a fresh college graduate, I entered Harvard as Barghoorn's student.

My goal at Harvard was to expand my undergraduate thesis into a Ph.D. dissertation on the Gunflint microfossils. During my first year, however, I got sidetracked working with Barghoorn and Warren G. Meischein, a world-class organic geochemist then employed at the Esso Research Laboratories in Linden, New Jersey, on a broad paleobiologic study of a 1,000-Ma-old shale deposit in northern Michigan. For me this turned out to be a wonderful education, but it gobbled up all of my time until our long manuscript was finally completed in August 1964. With the academic year about to begin, Barghoorn's plan to complete a second manuscript that summer, for the first time describing in detail the Gunflint microfossils, was put on hold.

Nothing of note had been published on the Gunflint fossils since the announcement of their discovery a decade earlier, in 1954, but in the interim much had transpired. Tyler's pioneering work had continued apace, and with support pouring in from the National Science Foundation he had employed several students in his laboratory at the University of Wisconsin who worked tirelessly, finding and photographing the tiny microscopic fossils. Diverse types of microorganisms had turned up, all new to science and some quite bizarre.

By 1958, Tyler had put the finishing touches on a detailed description of the geologic setting, mineralogy, and paleoenvironment of the Gunflint deposit and had forwarded the text, along with photographs of the fossils, to Barghoorn, whose task it was to interpret, formally describe, and officially name the newly discovered life-forms. But during the mid-1950s Barghoorn's personal life had taken a turn for the worse, and in the years since 1958 he had been unable to carry out his part of the job. Then, tragically, Tyler fell ill, and at the age of 57, in October 1963, he died, never to see the ripened fruits of his labor reach the printed page. Barghoorn traveled to Madison, Wisconsin, and brought back to Harvard Tyler's field notes, rock specimens, lab records, thin sections, and a huge number of photographs. I was given the task of sorting through this maze, of organizing the research materials of a senior scientist whom I much admired but had never met.

"What Are We Going to Do?"

At the end of the summer of 1964, Tyler's draft of the Gunflint manuscript still sat dusty and uncompleted, now after some six years. One afternoon in early fall, Barghoorn came storming up to my desk and, thrusting a pile of papers into my hands, shouted, "Look at this! What are we going to do?!" What at first seemed a thick jumble of disorganized pages was in fact a hefty manuscript on the Gunflint fossils written by Preston Cloud.

In their initial 1954 report of the Gunflint discovery, Tyler and Barghoorn had craftily neglected to pinpoint the location of their find (noting only that it was "near Schreiber" in "southern Ontario"), but Cloud ferreted it out. (I learned later that Cloud had traveled to Schreiber, rented a boat, found the fossil locality, and was stranded on the outcrop for three days by a fierce Lake Superior storm—a forced sojourn during which he carried out the first truly excellent detailed geologic study of the local area.)

The manuscript Barghoorn shoved into my hands had been submitted to *Science*, whose editor, Philip Abelson, had sent it to Barghoorn for scientific review in preparation for its publication. Barghoorn was more than a little agitated. He was livid. After all these years, he was about to be scooped!

Fuming, Barghoorn scurried around trying to locate the shelved 1958 Tyler text. I was assigned the task of preparing the review of Cloud's paper (and though I took pains to do the most thorough job I could, I discovered later that Cloud—no doubt thoroughly annoyed by the situation—ignored every one of my suggestions). By the end of the next day, after I had finished writing the review, Barghoorn had come up with a plan of attack: we would hold Cloud's manuscript for a week or two and use that time to get the Tyler-Barghoorn paper in publishable shape. Then Barghoorn would call the *Science* editor, Abelson, explain this "serendipitous coincidence" of timing, and convince Abelson to publish the Tyler-Barghoorn paper first, thereby retaining for them their claim of scientific priority.

The next two weeks were unbelievably hectic. Barghoorn had no courses to teach that semester and, as instructed, I cut all my classes. Tyler's research materials were a shambles, and by that time I had managed to sort through only about a third of several huge piles. Moreover, I had found to my horror that though the photographs were keyed to individual thin sections, there was no information to pinpoint the specific location within each section (the microscope stage coordinates) where Tyler's students had found the pictured fossils. This meant that I would have to spend many weeks, maybe months, scanning section after section in order to relocate the photographed specimens. Worse than trying to find numerous needles in multiple haystacks, given our time constraint there was no way this could be done.

Clearly, we would have to rely on what we already had on hand. That meant that Barghoorn's interpretations and descriptions of the fossils, and the naming of them (that is, the taxonomy of the Gunflint assemblage), would have to be based on the best photographs available rather than on examination of the individual specimens. (This, I am sorry to say, was not good science. No one, ever, should interpret, describe, and name new organisms without studying carefully the specimens themselves. It's true that we were forced into this by unusual circumstances, but this was not the proper way to do this job.)

At the beginning of each day, I would rummage through the photos and pick out the "best" for any given size-shape category (thin filaments, fat filaments, small spheres, big spheres, umbrella-shaped forms, and so forth), and then within each category would pile together those that to me looked most alike. Barghoorn would then check and revise my groupings and begin writing the formal taxonomic descriptions. By this time, he would have begun to formulate an appropriate Greek or Latin name for each morphologically defined genus (for example, *Gunflintia*, for the slender filaments, named after the geologic deposit; and *Eoastrion*, "dawn little star," for radiating filament clusters), names that I was sent to try out on Leslie Garay, the orchidologist down the hall who was an expert on such matters.

Barghoorn was working feverishly, as was I, and when he left at the end of each day I had the night job of preparing high-quality prints of each of the photographs selected, pictures that I could then paste together to make up figures to accompany the final published manuscript.

After nearly two weeks of this frantic activity, the job was nearing completion. Barghoorn had finished writing his portion of the text and the formal descriptions of the various types of fossils, and though I had not yet completed the final figures I did have good prints of practically all the photos to be included. It was now time to call Abelson at *Science*. But before that, the matter of authorship of the paper had to be sorted out.

In scientific circles, the authors of the first published paper of any series of papers that report similar new finds are credited with the discovery, establishing for its authors scientific priority, a sort of intellectual "ownership" rather like that gained from patenting a new invention. In the same way, in any multiauthored scientific work, the order of authorship makes some difference because the first (primary, principal, or senior) author is assumed to have contributed most to the project and, thus, to deserve the lion's share of the credit.

In this case, Tyler had discovered the deposit, recognized its significance, and (with his students) done most of the scientific work. Clearly, Tyler deserved the majority of the credit and both he and Barghoorn had always assumed that this magnum opus, like their

1954 paper on the Gunflint find, would have Tyler as the first author, Barghoorn the second.

But Tyler was now deceased. Barghoorn broached the subject, wondering, "What should we do?" I knew what he wanted me to say, and as his student I certainly wanted to please him. Still, I believe my reply was honest. I argued that the paper was bound to raise questions, that some might doubt the authenticity of the Gunflint fossils, and that because Barghoorn had interpreted, described, and named the new fossil microorganisms he obviously would have to take the heat. Someone had to stand up and be counted, and Barghoorn was the only one who could assume that responsibility. He seemed pleased by my response.

With the Barghoorn-Tyler order of authorship settled, he then invited me to join them as the third (junior) author of the paper. I thought the world of Professor Barghoorn, and though he had been kind to me before, this gesture was unbelievable! Still, I knew that I simply had not contributed enough to deserve such an honor. Respectfully, I declined.

(Over the years since I have wondered somewhat wistfully how my career might have soared had I accepted Barghoorn's generosity. The Barghoorn-Tyler paper is a classic. For all of time it will probably stand as the most important article ever written in the field, and it certainly would have made an impact on my life had it been by Barghoorn, Tyler, and Schopf! But credit *matters* in science, and it would have been wrong for me to have authored a work to which I contributed so marginally. I do, however, deeply value Barghoorn's published acknowledgment to me in the paper "for assistance . . . in taxonomic description" and I take personal pride in his use of a couple of the photographs from my undergraduate honors thesis, pictures of the Gunflint fossils that he regarded as among the best available.)

All was now in place to contact Abelson at *Science*. Barghoorn invited me into his office and made the call. Immediately, he found out that Cloud, too, had recently called Abelson, inquiring about the fate of his presumably "lost manuscript" for which Cloud had yet to receive prepublication scientific reviews.

Barghoorn assured Abelson that the review (mine) was completed and would soon be sent, and he then gave Abelson the story: "As luck would have it, I have a manuscript ready to submit on the same material . . . you know, the one I've been working on for years. As you may recall, Tyler and I announced our discovery in *Science*—your journal [*sic!*]—ten years ago. It would be only fair for us to have priority . . . our paper ought to be published first."

Abelson was a remarkable man, a first-rate chemist and an exceptionally gifted geologic-biologic scientist (though a disappointing public speaker, as I first discovered when I journeyed to Cleveland to hear him in my undergraduate days). As editor of *Science*, he had taken this formerly run-of-the-mill journal to a position of international prominence. Moreover, he was well acquainted with, knew the personalities of, both Barghoorn and Cloud; he knew the history and importance of the Gunflint discoveries; and, in June 1961, at a small scientific meeting held at the Blue Meadow Lodge in the Blue Ridge Mountains of Virginia, he had heard Barghoorn publicly announce, in Cloud's presence, that the Tyler-Barghoorn Gunflint paper would "soon be ready."

Abelson wanted no part of what seemed to him a brewing donnybrook. He wisely withdrew from this potential mess by making a deal with Barghoorn: "You talk with Pres Cloud. Whatever you two work out will be OK with me."

I still stood beside Barghoorn's desk as he telephoned Cloud. "Hello, Pres. Received your manuscript for review. With a few minor changes, it looks just fine. But, hey, as luck would have it, my paper with Stanley is finally ready to go. Really! I have it in my hands right now. It goes out tomorrow." Cloud said: "Seeing is believing!" (and Barghoorn covered the receiver and repeated to me: "He says, seeing is believing"). "You know, Pres, Stanley and I worked for years on this—our paper ought to appear in *Science* first. What do you say?" Cloud said: "Seeing is believing!" (again, Barghoorn repeated to me: "He says, seeing is believing").

The Barghoorn-Tyler manuscript, accompanied by my yet unfinished plates, was mailed to Cloud the next morning. By the time we had heard from him a week or so later, the final draft and finished plates were ready to submit to *Science*. Cloud was more than a little miffed, but he swallowed his pride and okayed Barghoorn to publish first. The two papers appeared in *Science* in 1965, the first in early

February, "Microorganisms from the Gunflint Chert" by Barghoorn and Tyler, followed a few weeks later by Cloud's article, "Significance of the Gunflint (Precambrian) Microflora." Landmark papers they were!

(To an outsider, the Barghoorn-Cloud battle over who could scoop whom must appear unseemly, even tawdry. And, of course, it is. But science is done by real people, and it's much more competitive than one might expect. Most turf fights never come to light—or like this one lie hidden for decades—but though they don't happen every day, they are not too uncommon. Probably the most famous centered on "Darwinian" evolution. Darwin gathered his evidence as the naturalist on the voyage of H.M.S. *Beagle* and by 1840 had put the finishing touches on a manuscript unveiling his idea of the "struggle for existence . . . descent with modification." But he was loathe to submit his article for publication, and it was not until early in 1858, nearly 20 years later, that he finally was spurred to action by seeing the very same idea in a manuscript sent him by British naturalist Alfred Russel Wallace (1823–1913) in the hope that Darwin would forward it for publication. On July 1 of that year the two received equal billing, their ideas presented back-to-back at a special meeting of the Linnean Society in London. Yet we remember Darwin, not Wallace, partly because only a year later *On the Origin of Species* appeared—thick with supporting facts, observations, even examples from everyday life—but also because Darwin belonged to the scientific elite and was in London, the seat of scientific authority, while Wallace was far afield, naturalizing in the wilds of the Malay Archipelago.)

The Floodgates Open Full Bore

Though the 1954 Tyler-Barghoorn announcement of discovery of the Gunflint fossils had stirred little reaction, the article of 1965 generated enormous interest. The phones in the lab rang off the hooks. We were besieged by reporters. Barghoorn was even interviewed on Boston TV, in those days a major coup. Precambrian life had become big news!

Within weeks, Barghoorn had received numerous invitations from colleges and universities to speak on the new finds. There was too much for him to handle alone so a certain number of these invites trickled down to me. This was a treat. I hadn't before done much public lecturing and I was lucky to have this chance. Moreover, though I was only a second-year graduate student I was gaining terrific exposure, making contacts that were certain to come in handy when it became time to look for a job.

Soon after I hit the lecture circuit, however, I was introduced to a reality I had not expected. Despite the evidence now amassed, the two landmark papers (not to mention my own honor's thesis, which hardly anyone knew about) and the undoubted prestige of Barghoorn, Cloud, *Science*, and Harvard, I was shocked to see the rampant skepticism. Bill Chaloner was right—this was a field in which "prudent caution" was very much required!

Dawson's debacle, the *Cryptozoon* controversy, Seward's criticism—all these were object lessons that had been handed down from professor to student, generation to generation, and all had become part of accepted academic lore. Barghoorn had known this for years (a factor, I now realized, underlying his reluctance to get the Gunflint project off the back burners), and it was now my turn to learn it firsthand.

The big problem was that the Gunflint organisms stood alone. They were isolated in time, seemingly marooned in the remote Precambrian, removed by nearly a billion and a half years from all other fossils known to science. Many of the old questions and quite a number of new ones came to the fore. Perhaps the deposit had been misdated and was not Precambrian at all. Perhaps the "fossils" were soil contaminants that had entered the rock through now-sealed cracks and crevices. Or maybe the tiny objects were simply needle- and ball-shaped mineral grains. Or bubbles in the rock. Or artifacts of thin-section preparation. Or some crystalline oddity associated with the formation of chert. Moreover, if *Cryptozoon* was "known" to be non-biologic, weren't the *Cryptozoon*-embedded filaments and spheres suspect too? Why hadn't cellular fossils turned up in other *Cryptozoon*-like structures, not even in the Phanerozoic?

Couldn't this whole business be some sort of fluke, some hugely embarrassing, awful mistake?

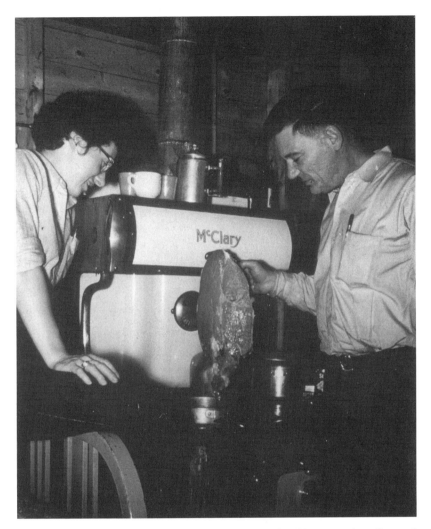

Figure 2.12 Elso and Dorothy (Osgood) Barghoorn, in 1964, preparing dinner after a long day's fieldwork collecting Gunflint chert.

A New Research Strategy Paves the Way

Fortunately for Barghoorn, Cloud, and me and my doctoral thesis (not to mention the future development of the science), these doubts soon could be laid to rest. During a field trip to Australia the previous year, Barghoorn had met an oil company geologist by the name of Helmut Wopfner who told him he remembered seeing beds of what might be

Figure 2.13 Black, coaly, flat-layered chert stromatolites of the richly fossiliferous 850-Ma-old Bitter Springs Formation of central Australia.

paleontologically promising Precambrian black chert in the vicinity of Alice Springs, on the northwestern margin of the Simpson Desert, deep in the Australian outback. At Wopfner's suggestion, Barghoorn and his wife-to-be, Dorothy Osgood, journeyed to the Ross River tourist camp east of Alice Springs, where, just as Wopfner had said, they found several 3- to 10-cm-thick beds of blocky weathering, wavy and thinly laminated black cherts associated with *Cryptozoon*-like structures from which they collected a few hand-sized specimens.

Once the Gunflint paper had been finished, Barghoorn turned the chunks of Australian rock over to me to see what I could find. I laboriously prepared half a dozen thin sections and was elated (to put it mildly) to discover they were full of microscopic fossils. There were many new types of filaments, spheres, and colonies of clustered cells, many that closely resembled living cyanobacteria (Walcott's "algae"). The new fossils were better preserved, *much* better preserved than those in the Gunflint rocks. I could see distinct cell walls, colony-encompassing mucilage sheaths, remnants of cell contents,

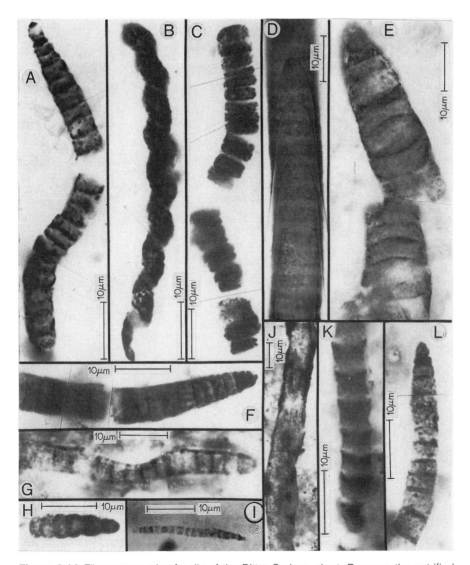

Figure 2.14 Filamentous microfossils of the Bitter Springs chert. Because the petrified microbes are three dimensional and sinuous, composite photos have been used to show the specimens in (A through G), (I), (K), and (L). (**A, F, I,** and **L**) *Cephalophytarion*; (**B**) *Helioconema*; (**C** and **G**) *Oscillatoriopsis*; (**D**) unnamed cyanobacterium; (**E**) *Obconicophycus*; (**H**) *Filiconstrictosus*; (**J**) *Siphonophycus*; (**K**) *Halythrix*.

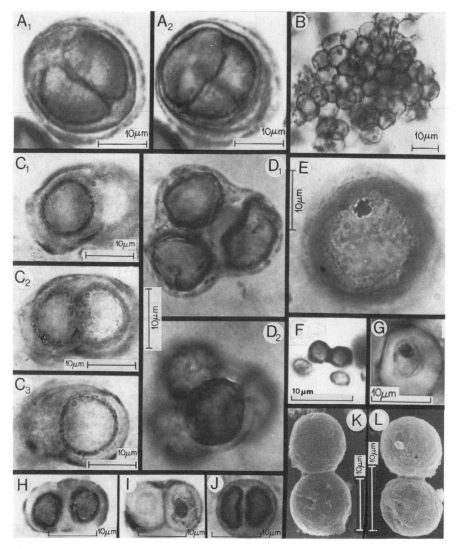

Figure 2.15 Optical photos (A through J) and scanning electron microscope pictures (K and L) of spheroidal microfossils from the Bitter Springs chert, in (A), (C), and (D) shown in multiple views of single specimens. (**A**) *Bigeminococcus*; (**B**) *Glenobotrydion*; (**C**, **H**, and **J**) *Eozygion*; (**D**) *Eotetrahedrion*; (**E**) *Globophycus*; (**F**) *Sphaerophycus*; (**G** and **I**) *Caryosphaeroides*; (**K** and **L**) unnamed paired cyanobacteria.

everything! For a few weeks I was the lab hero. And though the age of this Precambrian deposit (the Bitter Springs Formation) was known only approximately, it seemed likely that it was about 1,000 Ma, roughly half as old as the Gunflint chert.

With care I crafted a plate of photos and drafted a short manuscript—the very first scientific paper I put together all by myself. Barghoorn and I sent the paper off to *Science*. Its publication late that summer—now viewed in the context of the earlier landmark articles on the Gunflint organisms—heralded the birth of a new field of science.

This second find showed that the Gunflint fossils were no fluke. Indeed, probably the only truly odd thing about the Gunflint and Bitter Springs fossils is that similar deposits had not been discovered even earlier. Walcott and a few others in his time had started the train down the right track only for it to be derailed by Seward, Raymond, and skeptics of their ilk. But like Darwin, all had assumed the tried and true techniques of the Phanerozoic hunt for large fossils would prove equally rewarding in the Precambrian. Simply put, this was wrong.

The Gunflint and Bitter Springs articles of 1965 charted a new course, showing for the first time that a search strategy specifically focused on the peculiarities of the Precambrian fossil record would pay off. The four keys of the strategy were to search for (1) microscopic fossils in (2) black cherts that are (3) fine-grained and (4) associated with *Cryptozoon*-like structures. Each part was crucial.

1. Large, many-celled plants and animals, like those of the Phanerozoic, are now known not to have appeared until shortly before the beginning of the Cambrian. Except in immediately sub-Cambrian strata, the hunt for megascopic fossils in Precambrian deposits was doomed from the beginning!

2. The blackness of a chert commonly gives a good indication of its coaly, organic carbon content. Like abundantly fossiliferous coal deposits, cherts rich in petrified, organic-walled microfossils are usually a deep jet black color.

3. The fineness of the quartz grains making up a chert provides another hint of its fossil-bearing potential. Cherts subjected to the intense

heat and presssure of a mountain-building episode are composed of
large grains giving them a sugary appearance, whereas cherts that
have escaped fossil-destroying processes are made up of tiny grains
and have a waxy glasslike luster.

4. *Cryptozoon*-like structures (stromatolites) are now known to have
   been produced by microbial menageries, layer upon layer of microscopic organisms living together in localized ecological communities. Find these multilayered cabbagelike structures in the rock
   record, especially if they are composed of fine-grained black chert,
   and they are likely to contain fossilized remains of the microorganisms that built them.

Birth of a New Field of Science

Beginning in the mid-1960s and accelerating to the present, studies
of Precambrian life have boomed. Whether measured by time and
money, rocks studied, fossils found, articles published, new insights,
or public interest, the field has skyrocketed, culminating in recent
years with the discovery of the oldest fossils known—microscopic
cellular organisms nearly 3,500 Ma old, more than three-quarters the
age of the Earth.

As the field has soared, new strong pillars have been added to
its sound foundation. The Barghoorn-Tyler studies of *Cryptozoon*-
forming microbes have been expanded by work on living microbial
communities and of particular biochemicals (the nucleic acids of protein-manufacturing ribosomes) that place these microorganisms on
early branches of the Universal Tree of Life. Timofeev's pioneering
finds of fossil plankton in siltstones have been enlarged to reveal the
first extinctions in the history of life and the rapid rise of new biologic
types that accompanied the advent of sexual reproduction. Cloud's
benchmark environmental syntheses, strengthened by a new understanding of comparative planetology, atmospheric evolution, and
carbon isotope-based evidence of the history of photosynthesis, have
been sharpened and refined to reveal an increasingly focused picture
of the developing early Earth. And Glaessner's early studies of Precambrian animal fossils have grown into a global blizzard of activity
that has provided new insight into the classic Precambrian-Cambrian

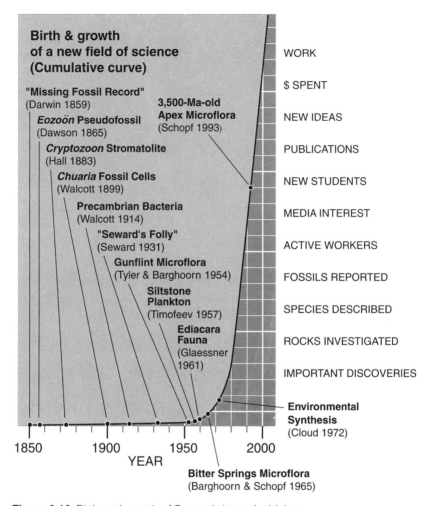

**Birth & growth
of a new field of science
(Cumulative curve)**

"Missing Fossil Record"
(Darwin 1859)

Eozoön Pseudofossil
(Dawson 1865)

Cryptozoon Stromatolite
(Hall 1883)

Chuaria Fossil Cells
(Walcott 1899)

Precambrian Bacteria
(Walcott 1914)

"Seward's Folly"
(Seward 1931)

Gunflint Microflora
(Tyler & Barghoorn 1954)

Siltstone
Plankton
(Timofeev 1957)

Ediacara
Fauna
(Glaessner
1961)

3,500-Ma-old
Apex Microflora
(Schopf 1993)

WORK

$ SPENT

NEW IDEAS

PUBLICATIONS

NEW STUDENTS

MEDIA INTEREST

ACTIVE WORKERS

FOSSILS REPORTED

SPECIES DESCRIBED

ROCKS INVESTIGATED

IMPORTANT DISCOVERIES

**Environmental
Synthesis**
(Cloud 1972)

1850   1900   1950   2000
YEAR

Bitter Springs Microflora
(Barghoorn & Schopf 1965)

Figure 2.16 Birth and growth of Precambrian paleobiology.

boundary problem: the previously mysterious abrupt rise of Phaner-ozoic multicelled life.

More than 99% of all scientists who have ever investigated the Precambrian fossil record are alive and working today. Discoveries are being reported at an ever-quickening clip, literally worldwide by workers in Australia, Brazil, Canada, China, England, France, Germany, India, Israel, Japan, Lithuania, Mexico, Russia, South Africa, Sweden, and the United States. This is not the place to list an honor roll of active Precambrian workers and their myriad achievements. I hope that I will be forgiven this omission but my colleagues will know, as do I, that this book is based on our collective contributions to the science and that all of us working today sit atop the broad shoulders of the few bold scientists who blazed this trail in the 1950s and 1960s, just as *their* course was set by the Dawsons, Walcotts, and Sewards, the pioneering pathfinders of the field.

The collective legacy of all who have played a role dates to Darwin and the dilemma of the missing Precambrian fossil record that he first posed. It is a great joy that after more than a century of trial and error, of search and final discovery, what was once "inexplicable" to him is no longer so to us.

The Oldest Fossils and
What They Mean

"Trust but Verify"

As we've seen in the first two chapters, the hunt for the oldest records of life is not a simple task. Like trying to solve a really complicated mystery when only a few clues have been revealed, it's easy to make mistakes. These can be worse than embarrassing, major blunders that set back the search for knowledge (like Dawson's "dawn animal" or "Seward's Folly"). Fortunately, science works by a system of checks and double-checks. The guiding rule, like that in international diplomacy, is to "trust but verify." And the verification must be by someone other than the discoverer (just as the Tyler-Barghoorn discovery was verified, and thereby certified, by Cloud). Because of these checks, research into the earliest records of life has been self-correcting—initial mistakes have been straightened out as experience and insight have been gained.

"Real World Problems" in the
Search for Early Life

With care and effort, errors in judgment and mistakes of misreading the record can and will be corrected. If there were nothing more to it, the hunt for remnants of ancient life would not be nearly so difficult. But the biology and the geology of the planet, factors over which we have no control, cause far greater problems.

Biology Causes Problems

Think about what happens when a hurricane or even a major windstorm rips through a neighborhood. Why don't limbs and trunks of dead trees litter the landscape for weeks or months thereafter? Usually they're carted off to a rubbish dump. But what happens to them there? They're buried. Then what happens? They're turned to sod. How? Worms, beetles, fungi, bacteria. Though organisms die around us every day, we are not knee-deep in carcasses. Individuals die, then decay to provide foodstuffs for hungry carcass feeders. The cycle of life, death, decay—followed by new life, new death, new decay— goes on.

Recycling of organic matter by the Earth's biology is very efficient. More than 99.9% of everything alive is eventually returned to some other living system. The small fraction left over, the less than 0.1% of dead organisms not recycled by the biology of the planet, is the pool from which fossils come.

Fossils are unrecycled bits and pieces of dead carcasses. Shells, bones, and other "hard parts" are resistant to decay, so we can understand how they can be preserved as fossils. But what about organisms that lack hard parts such as Precambrian cyanobacteria and simple single-celled algae? Their chances of being preserved are exceedingly slim. To be fossilized they have to be buried, usually rapidly, beyond the reach of the ravenous recyclers. Only a few ever make it into the rock record.

But there are other problems too. Because they lack hard parts, Precambrian microbes are easily crushed, so even if they escape recycling and are entombed in sediment they almost always are flattened beyond recognition by the compacting rock. And even if preserved in identifiable form, they are so small they are hard to detect. If we link these problems to what we know already—that Precambrian microorganisms can be discovered only if rare fossil-bearing rocks are collected in the field, prepared in the laboratory, and painstakingly studied by someone experienced enough to detect the tiny fossils and distinguish them from contaminants and fossil-like artifacts—it is a wonder that we have any meaningful early fossil record at all.

Geology Causes Even Greater Problems

Though the biology of the planet works against such finds, geologic constraints are worse. Geology is a double-edged sword. On the one hand, it is the deposition of sediments and their conversion to hard rocks that permits preservation of fossils in the first place. But on the other hand, as we have seen in chapter 1, because of ordinary geologic processes—mountain building, erosion, plate tectonics, and the like—only a small fraction of rocks that were once formed have survived to the present.

Like most things in nature, rock units have a "lifetime"—some exist only a short period, others much longer—on average, a few hundred million years. It's easy to see that young rocks should be common and old rocks rare. This is exactly the way it is. Phanerozoic rocks surviving to the present are abundant; those of the Proterozoic, less so; those of the Archean, rare; those as old as 3,500 Ma, the age of the oldest known fossils, nearly nonexistent. Though logically convoluted, "the older the rock, the greater the chance it no longer exists!" And if old rocks no longer exist, their entombed fossils have been lost as well.

There are other geologic constraints. The crust of the Earth, the outermost rind where fossil-bearing sediments are preserved, is a wafer-thin skin—only 0.3% of the volume of the planet. Temperature within the Earth increases by 20° to 30° C for each kilometer of depth. Burial to 10 kilometers or more produces rock temperatures of hundreds of degrees (accompanied by enormous pressures), and even shallower sediments are routinely heated to 150° C, enough to obliterate all but the most robust Precambrian organic-walled fossils. Because rock burial to fossil-frying depths happens more or less randomly over time, the odds of fossils being charred to obliteration increase with the age of the rock. In other words, the older the rock, the greater the chance its fossils have been grilled to nothingness.

Though past mistakes in the search for ancient life can be corrected, there is nothing we can do to change the biology and geology of the planet. At every turn the cards are stacked against the hunt, and the older the remnants sought, the worse the odds. Precambrian mi-

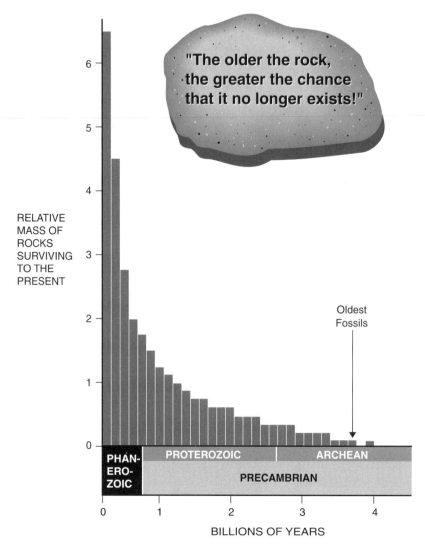

Figure 3.1 Survival of rocks over geologic time.

crobes lack hard parts, so most never get fossilized in an identifiable state. All are tiny, delicate, difficult to detect, easy to confuse with fossil-like artifacts. There are few places to search, especially in truly ancient terrains, because virtually all of the early rock record has been destroyed by geologic processes. And the sliver of potentially fossil-bearing sediments left is mostly so well cooked that any fossils once

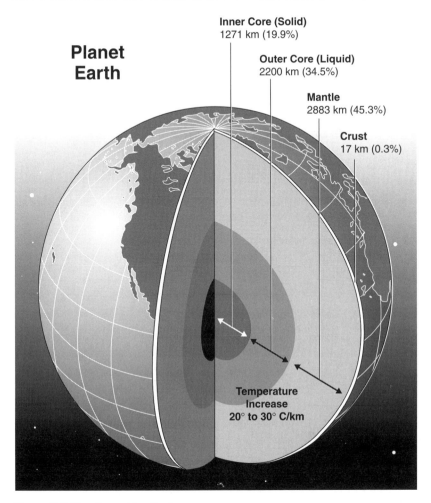

Figure 3.2 Structure of the Earth.

contained have been "fubarized"—fouled up beyond all recognition. Though this is a new science and the hunt has hardly begun, it already is clear that the early fossil record will yield its secrets only grudgingly.

Questions and Answers about the Oldest Records of Life

In 1992 and '93 I authored two articles reporting the discovery of the oldest fossils known to science. These microscopic threadlike organ-

SCIENCE • VOL. **260**: 640-646 • 30 APRIL 1993

■ RESEARCH ARTICLE ▬▬▬▬▬▬▬▬▬▬▬▬▬▬▬

Microfossils of the Early Archean Apex Chert: New Evidence of the Antiquity of Life

J. William Schopf

Eleven taxa (including eight heretofore undescribed species) of cellularly preserved filamentous microbes, among the oldest fossils known, have been discovered in a bedded chert unit of the Early Archean Apex Basalt of northwestern Western Australia. This prokaryotic assemblage establishes that trichomic cyanobacterium-like microorganisms were extant and morphologically diverse at least as early as ~3465 million years ago and suggests that oxygen-producing photoautotrophy may have already evolved by this early stage in biotic history.

Figure 3.3 Title page of the 1993 article on the oldest fossils known—published in the same journal and exactly 39 years to the day later than the groundbreaking Tyler-Barghoorn article of April 30, 1954.

isms, made up of arrays of tiny cells like beads on a string, are petrified in a 3,465-Ma-old rock unit of northwestern Western Australia known as the Apex chert.

Studies of this deposit have been a challenge. A team of colleagues and I collected the rocks in 1982, and the hunt for fossils began soon after in my laboratory at UCLA. One after the other, two graduate students found nothing. Then in 1986 a third found what eventually proved to be true fossils, but the student was more geologist than biologist and chose to do doctoral work on a rock-focused research topic. It was not until three years later, when I was a year-long visitor at London's Natural History Museum (then the "British Museum [Natural History]"), that I finally had a block of time to devote to the study. There I discovered good examples of three distinct kinds of microorganisms (reported in the 1992 article). After my return to UCLA, I set aside several months for microscope work and found eight more species (reported in 1993).

A guiding rule in science is that evidence for new claims must be made available for verification. In this instance, the evidence is the

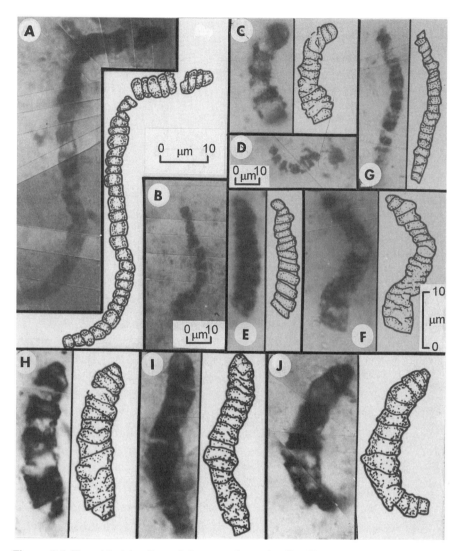

Figure 3.4 The oldest fossils, cellular cyanobacteriumlike filaments shown in thin sections of the 3,465 ± 5-Ma-old Apex chert of northwestern Western Australia. Because the petrified microbes are three dimensional and sinuous, composite photos have been used to show the specimens in (A), (B), (D), and (F through J). (**A** and **B**) *Primaevifilum amoenum*, a species having cylinder-shaped body cells and rounded end cells; (**C** through **F**) *Archaeoscillatoriopsis disciformis*, tapered filaments composed of disk-shaped body cells; (**G**) *Primaevifilum delicatulum*, a narrow species of the genus *Primaevifilum*; (**H**, **I**, and **J**) *Primaevifilum conicoterminatum*, a species that has distinctive conical end cells. [The 10-μm-long scale in (F) shows the magnification of all fossils except as otherwise indicated in (A), (B), and (D).]

fossils themselves, so the named specimens (technically, "holotypes") are archived at the museum in London, where they are open to study by others. Backup specimens are in the collections of the Geological Survey of Western Australia, which provided field help when the fossil-bearing rocks were collected.

The Apex fossils are scrappy. Hard to find. Difficult to study. They are abundant but charred, shredded, overly cooked. Tiny bits and pieces are common but generally nondescript; short two- or three-celled fragments are rare and easy to overlook; many-celled specimens are few and far between; and fossils that could be called "well-preserved"—like those of the Gunflint and Bitter Springs deposits—are nonexistent. Were these remnants not so remarkably ancient they would not merit much attention.

It seems to me likely that several of the Apex species are cyanobacteria, a fairly advanced group of microorganisms that until this find was not guessed to be present so early in Earth history. But whether they are cyanobacteria or not, no one disputes their importance as "real fossils" of very ancient forms of life. Their discovery tells us new things about life's early history, most notably that life not only existed but was already flourishing only a few hundred million years after the Earth first became habitable. The find sets a firm date on how long life has been present on the planet, a benchmark recounted worldwide in textbooks, encyclopedias, the popular press (*Time*, *Newsweek*, newspapers, television), even in the *Guinness Book of World Records*.

Since discovery of these "oldest known fossils" I have given a number of public talks in which they were featured. Time after time the same questions are raised, some a bit technical, most less so. How were the fossils discovered? How can we be sure they are actually remnants of life? What is known (and what is not) about these tiny ancient organisms? Why do they matter? Together, these questions get to the heart of what these fossils mean, and their answers serve as a case study that shows how this type of science is done. The following are the "Top Ten" most often asked questions about this remarkably ancient find.

1. How do you know where to hunt for ancient fossils?

At some point, usually early, in the history of a country, it becomes important to inventory its natural resources. Are there any diamonds to be found? What about gold, iron, copper, oil, coal? In the United States this is the job of the U.S. Geological Survey, the organization that was headed by C. D. Walcott in the early 1900s. In Australia the task is shared by the federal Bureau of Mineral Resources (recently renamed the Australian Geological Survey Organisation) and various state agencies such as the Geological Survey of Western Australia.

The charge of geological surveys is twofold. First, to explore the land, identify rock strata (especially those of practical value), and make maps showing the rocks' geographic spread. Because these geologic maps are paramount to a nation's economy—and critical to the defense industry as well—they have been treated in some countries as state secrets, a policy that in the past has slowed my fieldwork both in China and Russia. Today such secrecy is rare since much geologic mapping is done via satellites from space, and not only the satellite photos but the maps produced are available for purchase by anyone, anywhere. (On two different trips in the early 1980s, my fieldwork east of Beijing and just south of China's Great Wall was thwarted by lack of geologic maps. I finally rectified this by climbing a mountain and, with binoculars in hand, made my own. On my next visit I presented a gift to the village elders: a large satellite photo I had purchased in the U.S. showing their commune with its fields, farmhouses, and the nearby Great Wall.)

Second, geological surveys determine the ages of the rock units mapped. These ages matter because certain types of economically important rocks were formed abundantly only during particular times of Earth history. For instance, iron-rich rocks (for making steel) and certain types of uranium deposits (for nuclear reactors) are plentiful only in Precambrian strata older than about 2,000 Ma, whereas coal and oil reserves are common in particular parts of the Phanerozoic. Age determinations also help pinpoint the time ranges of the index fossils used to match up strata from one place to another, as we saw in Chapter 1, and are crucial to figuring out the speed of evolution. And

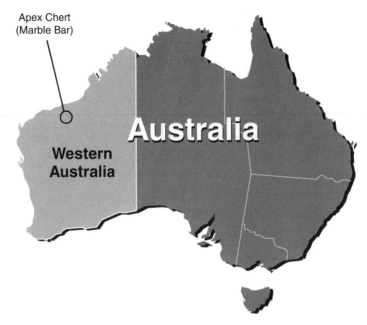

Apex Chert
(Marble Bar)

Western
Australia

Australia

Figure 3.5 Location of the Marble Bar site of the Apex chert in northwestern Western Australia.

because the times spanned by Precambrian iron and uranium ores were limited by changes in the environment caused by evolving life, knowing the ages of these deposits helps unravel the early histories of life and the global climate.

To find the oldest fossils, exceptionally old (Archean) rocks must first be located. But the fossils will have survived only if the rocks are relatively uncooked, little metamorphosed. Only two terrains in the entire world fill the bill, both made up of strata 3,000 to 3,500 Ma old. (1) A thick pile of volcanic and sedimentary rocks known as the Pilbara Supergroup (from an area known geologically as the Pilbara Craton), west of the Great Sandy Desert near the northwesternmost corner of Australia. (2) A sequence of rocks in eastern South Africa known as the Swaziland Supergroup that forms the hills and valleys of the Barberton Mountain Land in and near the kingdom of Swaziland. Both are somewhat metamorphosed, but the Australian rocks, the source of the oldest known fossils, are less pressure cooked.

Fieldwork in the Australian Pilbara is demanding. The hub of the area is Port Hedland at the northwestern edge of the country, a coastal town of a few thousand inhabitants. From here, Precambrian (2,500-Ma-old) iron ore dug up at Mt. Tom Price, the world's largest open-pit iron mine about 300 kilometers to the south, is shipped to steel mills throughout the world, mostly to Japan, where it is forged into girders and bridge spans, Toyotas, Hondas, and the like. Away from Port Hedland and the company settlements near the iron mines, the silent outback is desolate. Homesteads can be hundreds of square kilometers in area. There are so few inhabitants that schooling is done by two-way shortwave radio. The landscape is a hilly, dusty desert, pockmarked by gullies and ravines and laced by dry broad riverbeds with little vegetation except for *Spinafex*, an Australian grass filled with sharp silica needles that tear at your clothes. There are almost no paved roads, hardly any towns, and even the watering holes are few.

The region can be traversed on foot, horsesback, or via the geologist's preferred vehicle, a four-wheel-drive Land Rover (equipped with camping gear, plenty of dried food, two-way radios, extra petrol tanks, and stout, front-end "roo-bars" to fend off roaming kangaroos). Fieldwork is carried out from a base camp, usually along a dry, sandy stream course where local gum trees (*Eucalyptus*) provide relief from the scorching sun. Populated only by an occasional gold prospector (all of whom seem to be wrinkled, bearded, overweight characters leading lumbering pack mules, just like in the movies) and even rarer emus, kangaroos, scrawny dingos, and stray cattle, it's a startlingly still country with great geologic exposures.

Iron, uranium, gold, and barite (barium sulfate, $BaSO_4$, a dense mineral used in driller's mud) are mined in the region, and because of the area's economic importance its geology has been mapped in detail, chiefly by field parties of the Geological Survey of Western Australia. Their published maps, geologic descriptions, and information on the age of the Pilbara strata paved the way for discovery of the Apex fossils.

2. How can you tell whether a rock is likely to contain microscopic fossils?

The basic search strategy has remained unchanged since it was developed in the mid-1960s: look in black (carbon-rich) cherts that are fine grained (unmetamorphosed) and associated with *Cryptozoon*-like stromatolites. Because of the way stromatolitic rocks form, most lack fossils. It's only the cherty ones that are promising, and only the fine-grained black cherts that are likely to contain fossil cells.

As we'll see later (chapter 7), stromatolites are formed by stacked, thinly layered matlike communities of microscopic organisms, most often in settings where limestones are laid down. But as the limestone-forming muds solidify to rock, the grains of calcite (calcium carbonate, $CaCO_3$) that make them up slowly enlarge and crush any microbes trapped between them. Hardly any cellular fossils have been found in limestone stromatolites. On the other hand, chert stromatolites and chertified parts of limestone stromatolites often harbor fossil cells. Fossil-bearing cherts are made up of tiny interlocking grains of quartz laid down from solution. The precipitated grains initially pass through a gemlike opaline state, taking thousands of years to solidify into a full-fledged chert, and the tiny microorganisms are petrified (technically, "permineralized"), embedded within a solid chunk of rock. The quartz grains, which are deposited inside the cells and surround them on all sides, develop so slowly they grow through the cell walls instead of crushing them. As a result, the petrified fossils are preserved in three dimensions—unflattened bodies that, except for their quartz-filled interiors and the brownish color of the aged organic matter that makes them up, bear a striking resemblance to present-day microbes (see plates 1 and 2).

In the Proterozoic Precambrian, stromatolites are common and serve as useful markers in the search for remnants of life. In the Australian Pilbara, however, as in most Archean terrains, these large, easily recognized microbe-built structures are rare, so the search strategy must focus instead on carbon-rich, fine-grained cherts. Unfortunately, there is no way to know in advance whether a particular chert, even a stromatolitic one, harbors fossils. I have been disappointed many times by rocks that "looked good" at an outcrop but on microscopic study turned out to be too cooked to contain identifiable cells.

Fossils in younger, Phanerozoic rocks are large and often easy to find in the field. Except for stromatolites (which evidence the presence of life, even though they are not themselves fossils of individual organisms), this is not so in the Precambrian. The hunt for tiny cellular Precambrian fossils takes place in the laboratory and demands long periods of painstaking microscopic search.

3. What signs of ancient life do you search for?

There are three independent, yet mutually reinforcing, lines of evidence: stromatolites, cellular fossils, and biologically produced coaly organic matter.

Stromatolites, large concentrically layered *Cryptozoon*-like structures, tell us that communities of microscopic organisms were present. Their size, shape, relation to surrounding sediments, and mineral makeup (studied in petrographic thin sections) say much about the local environment. But because most stromatolites do not contain cellularly preserved fossils, they usually reveal little about the microorganisms that built them.

Cellular fossils, studied with a high-powered microscope in petrographic thin sections or acid-resistant residues, often show practically the same features of size, shape, cellular structure, and colony form as microorganisms living today. Their comparison with modern microbes can provide firm insights about early evolution and the makeup of the ancient living world. But their preservation requires geologically rather mild conditions, so they are found rarely in rocks heated above 150° C, and never in deposits severely pressure cooked.

The chemical (isotopic) composition of the carbon in the particulate, coaly, organic matter ("kerogen") of ancient sediments tells us whether photosynthetic organisms were present where the rocks were formed, a telltale signature that can survive even if cellular fossils are destroyed. Other chemical elements, for instance sulfur, can be used in a similar way. As we will see in chapter 6, these signposts of ancient life are especially interesting since they show how early microbes lived.

4. How do you know exactly where these rocks came from?

Though this is a simple, even trivial matter, it is surprisingly important. An object lesson underscores the point.

Early in the 1980s, two others and I reported the discovery of microscopic fossils in cherts from strata known as the Warrawoona

Group in another region of the Pilbara Craton (about 45 kilometers northwest of the tiny town of Marble Bar near where the fossil-bearing Apex cherts were later collected). These Warrawoona cherts were from the "North Pole Dome area," so dubbed by early miners who thought this site was as isolated from civilization as the North Pole.

My part in the project was to detect, describe, and name the new finds based on thin-section study of rocks that had been collected by one of the others in 1977. Because of bad weather the collection had been made in haste, and the collector had traveled alone using only road maps rather than detailed geologic maps or aerial photos of the region. After laboratory work was completed, he revisited the site to collect more samples. To his chagrin he was unable to locate the specific bed he had sampled before. Others, myself included, have several times tried to refind this unit, but no one has been successful.

The rule in science is trust but verify. But if a fossil-bearing bed cannot be relocated, the find cannot be confirmed. I have no doubt that the microscopic objects we reported from this lost rock unit are authentic fossils, but I now know we should have insisted that their source be proven before the find was announced.

As the old saying goes, "Once fooled, shame on you. Twice fooled, shame on me." I took special pains to avoid a repeat of the Warrawoona fiasco when in 1986 fossils were first found in the Apex chert. I pulled out my field notes and photos from the expedition during which our team had collected the rocks; pinpointed the locality on aerial photos and maps, both geologic and topographic; and handed these over to a UCLA geology student who was heading off for fieldwork in northwestern Australia. Thus armed, the student collected samples from the exact rock bed at the same locality. Work in the lab showed these cherts to contain fossils identical to those in the original samples. We announced discovery of the Apex fossils only after this confirmation.

Replicate sampling—now by other geologists and paleontologists as well (not to mention Australian, American, British, and Japanese television crews)—documents the source of these oldest-known fossil-bearing rocks.

Figure 3.6 Geologist Kase Klein (*right*, of the University of New Mexico) and biologist David J. Chapman (of the University of California, Santa Barbara) collecting samples of Apex chert at the Marble Bar locality. (Courtesy of J. M. Hayes, Woods Hole Oceanographic Institution, Massachusetts; photo by J.C.G. Walker, University of Michigan.)

5. How can you know the age of the Apex fossils so precisely?

The Apex fossils are composed of coaly, carbon-rich organic matter. Because dating of fossil humans and human artifacts is often done using the radioactive isotope of carbon, ^{14}C, it might be thought that the Apex fossils were dated by the same technique. Not so. Radioactive carbon spontaneously decays to a stable isotopic form of nitrogen, ^{14}N, so rapidly that after 50,000 to 60,000 years too little remains to be detected by even the most advanced equipment. Rad-

Figure 3.7 An outcropping of the Apex chert at Marble Bar. (Courtesy of J. M. Hayes, Woods Hole Oceanographic Institution, Massachusetts; photo by J.C.G. Walker, University of Michigan.)

iocarbon can be used to date organic remains younger than about 60,000 years but is useless for anything older.

The Apex fossils are billions, rather than only tens of thousands of years old, so ^{14}C dating can't be used. But other isotopes can—those that decay exceedingly slowly. The rate of radioactive decay can be determined by laboratory experiment, and once this rate is known one only has to compare the amount of the original (parent) isotope that remains undecayed to the amount of the new (daughter) isotope that has been formed, and then calculate the length of time this has taken.

For example, imagine that radioactive isotope "X" spontaneously decays to produce isotope "Y" and that the rate of this change (called

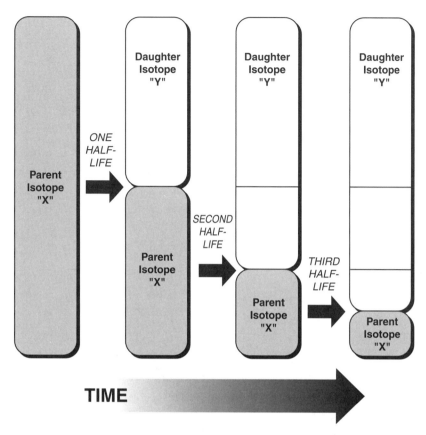

Figure 3.8 In radioactive decay, if parent isotope "X" decays to daughter isotope "Y," then, after a time known as one "half-life," 50% of the parent will have decayed to form an equal number of atoms of the daughter. If the half-life of radioactive parent X is known, the age of a mineral containing X and its daughter isotope Y can be determined from the ratio of parent to daughter.

the half-life) is such that it takes 1,000 million years for half of a particular amount of X to be converted into newly formed Y. If we then determine that a sample contains one gram of undecayed X and one gram of newly formed Y, it is easy to calculate that it originally contained two grams of X (half subsequently converted to Y) and that the decay took 1,000 million years, the age of the sample studied.

This dating technique needs a sensitive (and expensive) mass spectrometer to measure amounts of the isotopes and also requires that both the undecayed parent isotope and the produced daughter isotope are trapped in the material examined. If either has escaped over time, the calculated age will be wrong. Crystalline minerals provide the permanently sealed trapping system needed. The best of these are minerals known as zircons, tightly sealed crystals that form when molten lava solidifies on the flanks of a volcano. As zircon grains crystallize, they seal in radioactive uranium (specifically, ^{238}U), which over time and through many steps decays to produce lead (^{206}Pb).

The rate of decay of ^{238}U to ^{206}Pb is incredibly slow, having a half-life of about 4,500 million years. In other words, if we had a sample of lava that survived from when the Earth was born, only half of the ^{238}U originally trapped in its zircons would by today be converted to ^{206}Pb. Because the U-Pb zircon technique permits accurate measurement of exceedingly ancient ages it is especially good for dating Precambrian rocks.

The Apex fossils are preserved in a chert bed sandwiched between two massive lavas of the Pilbara sequence. Because of its relation to these lavas it has been dated quite precisely. Zircons in a lava immediately overlying the fossil-bearing chert have a U-Pb age 3,458 ± 1.9 Ma, whereas those in a lava below the chert are 3,471 ± 5 Ma in age. The fossil-bearing rock unit is therefore older than about 3,460 Ma and younger than about 3,470 Ma; the bed has an age of 3,465 ± 5 Ma.

This precision is wonderful. A decade ago we would have been forced to be satisfied with an age estimate ± 10%. For fossils 3,465 Ma old, this 10% uncertainty (that is, ± 347 Ma) is equivalent to a range of some 700 Ma (from about 3,100 to 3,800 Ma), a time span longer than the entire Phanerozoic! Geochronologists, the specialists who carry out this work, have done a superb job. Thanks to them we now have a technique for the precise dating of very ancient rocks.

Interestingly, though the rock unit that contains the Apex fossils has an age of 3,465 ± 5 Ma, the fossils themselves are actually even older. The fossils are preserved in small rounded granules, a millimeter or so in size, that are embedded in a type of rock known as a grainstone conglomerate and made up of many such rocky chunks. Conglomerates like the Apex chert are formed along wave-washed beaches and at the mouths of streams and rivers. The granules and pebbles that make up the deposit are bits and pieces of rocks origi-

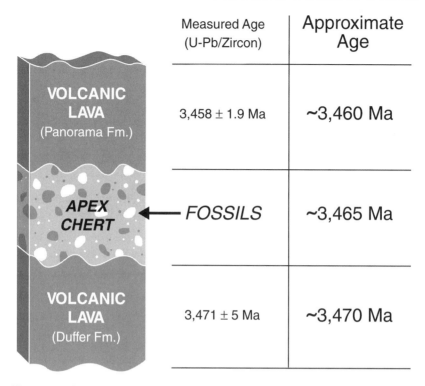

	Measured Age (U-Pb/Zircon)	Approximate Age
VOLCANIC LAVA (Panorama Fm.)	3,458 ± 1.9 Ma	~3,460 Ma
APEX CHERT ← FOSSILS		~3,465 Ma
VOLCANIC LAVA (Duffer Fm.)	3,471 ± 5 Ma	~3,470 Ma

Figure 3.9 Sandwiched between two precisely dated volcanic lava beds, the Apex chert is 3,465 ± 5 Ma old.

nally formed someplace else and then broken up and carried by flowing water to the Apex bed. Only one of the many types of rounded stones in the conglomerate contains fossils. Their well-rounded sides and small size shows that they were transported a long distance, but no one knows from where. Unless the "mother load," the source bed, can be found and dated, we will never know the full age of the Apex fossils. They are older than 3,465 ± 5 Ma, but how much older remains a question.

6. What was the Earth like when the Apex organisms were alive?

The geology of the Pilbara Craton provides good clues. The total rock sequence is massive, some 10 to 15 kilometers thick, in places lying nearly flat like the layers of a cake but elsewhere it is gently tilted up by past crustal movements. Because of the arid climate and lack of

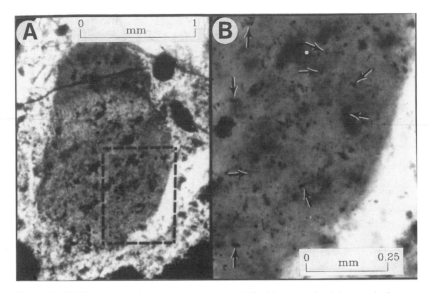

Figure 3.10 The Apex microorganisms are petrified in organic-rich rounded granules (**A**) in which, as shown at higher magnification in (**B**), they are randomly scattered (at arrows) like raisins in a loaf of raisin bread.

plant cover the rocks are exceptionally well exposed to view, and though over time they have been somewhat pressure-cooked, they are less altered than any other similarly aged sequence.

Overall, the region is composed of large, low, egg-shaped granite hills (technically, "batholiths"), 20 to 50 kilometers in diameter, each surrounded by thick piles of interlayered volcanic and sedimentary rocks called "greenstone belts." The geology of these belts tells us they were laid down in shallow seas. Some contain bedded sulfate deposits (composed originally of the mineral gypsum, hydrous calcium sulfate, $CaSO_4 \cdot 2H_2O$), formed when seawater lagoons evaporated billions of years ago. Others have silts and sands layered in ways that could have formed only in very shallow basins. Still others contain rocks broken up and jumbled by storm-driven waves, a sure sign of a shallow water setting, and "pillow lavas," mound-shaped masses formed when volcanic lavas ooze into shallow seas.

The scene was dominated by broad, shallow seaways into which volcanic lavas erupted. Scattered volcanic islands were fringed by river gravels, sandy inlets, mudflats, and occasional evaporitic lagoons.

Figure 3.11 A reconstruction of the 3,500-Ma-old environmental setting of the Apex chert displayed at the U.S. Natural History Museum, Washington, D.C., and designed by K. M. Towe of the Smithsonian Institution. (Courtesy of K. M. Towe.)

The Apex chert occurs within one such shallow-water sequence, sandwiched between two massive lava flows, on the western flank of what is known as the Mount Edgar Batholith.

The Australian Pilbara and the South African Swaziland rocks, the only thick geologic sequences known to have survived from this distant time, are made up mostly of greenstone belts. It is of course chancy to guess at the geology of the entire globe based on these two small areas. But as the only remnants we have to go on, they paint a picture of wide shallow seas dotted with abundant rocky volcanic islands and their smoldering fumaroles and hot springs.

On a broader scale, we can surmise that whatever continents existed were small by present-day standards and that they trucked across the global surface faster than today, powered by the large amounts of heat escaping from the Earth's interior. Because the Moon was closer than at present, the Earth revolved more rapidly and days were shorter, tides greater, storms more severe. The skies were a hazy steel-gray blue, darkened by dust storms, volcanic clouds, and fine rocky debris kicked up by bombarding meteorites. The atmosphere was rich in nitrogen, carbon dioxide, and water vapor but contained only trace amounts of oxygen gas (O_2). Though no one knows for sure, the early oceans may have been shallower than those today, and atmospheric carbon dioxide (CO_2) seems to have been present in amounts large enough to produce greenhouse warming and perhaps a tropical or even hotter climate. Because of the near absence of free oxygen, ultraviolet-absorbing atmospheric ozone (O_3) was in short supply, so the Earth's surface was bathed in ultraviolet (UV) light lethal to early life. Organisms had to learn to cope with this harsh environment, and the rocky UV-drenched landscape would have remained lifeless until biochemical mechanisms evolved to protect cells from UV rays and repair the damage it causes.

7. How can you prove that the fossils are actually as old as the rocks?

It is crucial to show that tiny objects claimed to be fossils are part of (that is, indigenous to) and as old as (syngenetic with the deposition of) the rock in which they are said to be present. This is especially important for claims of exceedingly ancient fossils, because if such claims are true they extend the frontiers of knowledge and are likely to be trumpeted in the press and find their way into textbooks. Any mistake would be a disaster!

In years past, as the search for Precambrian life was getting underway, contamination was a serious problem. Microscopic organisms are everywhere—in the air we breathe and the water we drink, on our clothes, furniture, and laboratory benches, even on our hair and skin—and if they somehow get into samples being studied they can be confused easily with true fossils. Fortunately, as the field has matured hard-and-fast rules have been laid down that all but eliminate repeating past errors. In particular, use of petrographic thin sections rules out contamination from acids and laboratory water during sample preparation. In such sections, only those objects that are entirely entombed in rock can be considered fossil, so it is easy to exclude contaminants that settle onto the surface of a section or are embedded in the resin used to cement the sliver of rock onto the glass thin-section mount.

Thin sections also provide a way to show that fossil-like objects date from the time a rock formed rather than having been sealed later

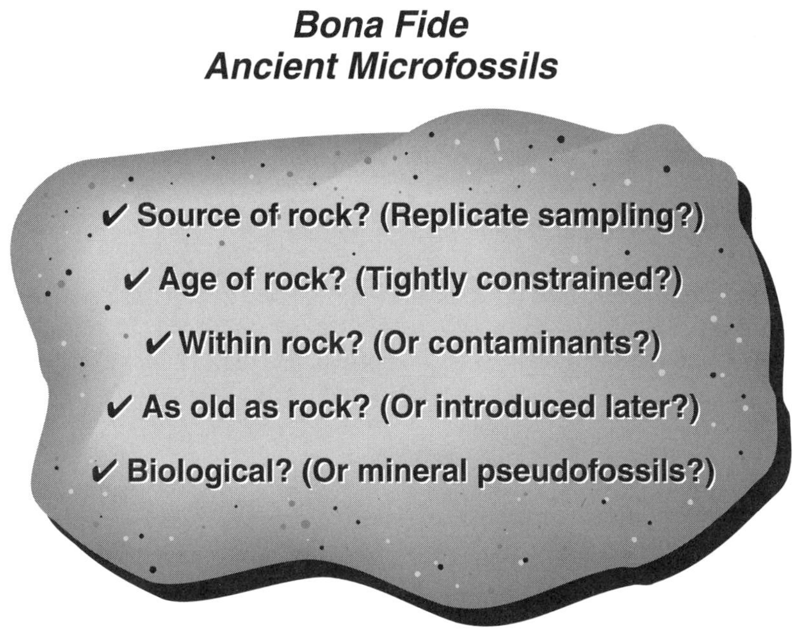

Bona Fide
Ancient Microfossils

✔ Source of rock? (Replicate sampling?)

✔ Age of rock? (Tightly constrained?)

✔ Within rock? (Or contaminants?)

✔ As old as rock? (Or introduced later?)

✔ Biological? (Or mineral pseudofossils?)

Figure 3.12 Tests that must be met by bona fide ancient microfossils.

in cracks and crevices. When quartz grains of a chert crystallize, they grow slowly against one another to form an interlocking network, a pattern in three dimensions like the way irregularly shaped rocks of a flagstone pavement fit into one another. But proving that a fossil is embedded in chert is not enough, for when cherts first form, many contain holes ("vugs"), especially those formed in stromatolites where gases (often oxygen, carbon dioxide, hydrogen, or methane) given off by the microbial community build up in small pockets. Later these cavities can be sealed, filled by a second generation of quartz laid down from seeping groundwater, sometimes tens or hundreds of millions of years after the first chert formed. Microscopic organisms trapped in these cavities and petrified by the second-generation quartz would be "good fossils" but would be far younger than the rock unit itself. Fortunately, the various generations of quartz in a chert are easy to sort out. Rather than having interlocking grains, quartz that fills cavities is a type known as "chalcedony"; it follows the smooth contours of the infilled pocket to form distinctive grapelike (botryoidal)

masses. Secondary quartz in cracks or veinlets is also easy to identify since it is angular and its grains are much larger than those first
formed.

Because special equipment is needed to prepare thin sections—and
their study is exceedingly time consuming—some workers have focused their hunt for ancient fossils on acid-resistant rock residues. In
the Proterozoic, where the fossil record is now well enough known
that misidentification of contaminants and fossil-like artifacts can be
avoided, this technique is useful, simple, and fast. In the Archean,
however, the fossil record is still largely a mystery. To avoid mistakes,
use of the more rigorous thin-section technique is essential.

All of the fossils reported from the Apex chert have been detected,
studied, and photographed in thin sections. They definitely are part of
the rock, wholly embedded in grains of first-generation interlocking
quartz.

8. How do you know the Apex objects are fossils, not mineralic look-alikes?

So little is known about Archean life that almost any new find is
likely to seem important. This makes it crucial to avoid mistakes. The
rules of the search must be even more demanding than in the hunt for
Proterozoic life, where the record is better documented and mistakes
are less likely.

In the past, microstructures "unlike known mineral forms" have
been said to be "Archean fossils" simply for want of any other explanation. Living contaminants, "lifelike" dust particles, ball-shaped
mineral grains, clumps and shreds of compressed coaly organic matter, solid opaque globules, and a variety of other objects have all been
claimed as Archean fossils, often on the basis of only one or a few
specimens and despite the absence of identifiable cells or other telltale
features of living systems. Many of these reports are founded on the
notion that because an object doesn't look mineral it "must" be fossil.
This negative reasoning, an error in logic, is not good enough. The
earliest records of life need to be backed up by positive evidence,
hard facts showing what an object *is* rather than what it seemingly is
not. And this positive evidence must be strong enough to rule out
plausible nonbiologic sources.

For example, because organic matter can be produced in nonbiologic ways (as when life originated, or today in interstellar space, as we will see in chapter 4), the mere presence of coaly particles in an Archean sediment is not enough to prove that life existed. And because uni-cell-like organic spheroids can form without life (from clumping of organic matter in seawater or by coaly matter coating ball-shaped mineral grains), tiny round organic bodies in a rock cannot be regarded as assured fossil cells. Nit-picking care of this sort is no longer necessary for reports of fossils from the Proterozoic where the evidence of life is overwhelming. But in the Archean, where so little is yet known, demanding rules must still apply.

Probably the best way to avoid being fooled by nonbiologic structures is to accept as bona fide fossils only those of fairly complex form. For truly ancient fossils this may seem an unreasonably stringent rule since the earliest kinds of cellular life almost certainly were very simple—probably individual tiny balls. But this is a young science, and until we have a sounder base of knowledge and better rules to separate nonfossils from true ones it's best to err on the side of caution. If we want to be certain, we must be careful. For the present, it is safest to accept only those fossils that have unquestionably biologic form, for example colonies of ball-shaped cells embedded in a surrounding organic envelope and threadlike filaments made up of chains of many cells. As evidence builds, we will gain confidence to better interpret the less certain finds.

The biologic origin of claimed Archean fossils can be accepted if they are (1) made up of coaly organic matter; (2) complex enough in cellular structure to rule out plausible nonbiologic sources; and (3) represented by numerous specimens (if one fossil can be preserved, others should be too). Like younger fossils and living microbes, they also should (4) be part of a once-living species population (with its gene-based range of differences among the members); (5) inhabit a livable environment; (6) grow and reproduce by biologic ways of cell division; and (7), if photosynthetic, exhibit the carbon isotopic signature of photosynthesis.

The eleven types of fossils in the Apex chert meet these tests. They are (1) composed of dark brown to nearly opaque (carbonized) coaly organic matter; (2) unlike nonbiologic structures and much more complicated: sinuous fossil filaments made up of cylinder-, box-, disk-, or barrel-shaped body cells and flat, rounded, muffin-, or cone-shaped end cells; and (3) known from a large number of specimens (nearly 1,900 cells measured in some two hundred individuals). Moreover, the Apex species (4) are each made up of members that vary slightly in size and shape, like the individuals in species of younger microbial fossils and living microorganisms; (5) were bottom dwelling, shielded from harmful UV light by overlying water and thick surrounding mucilage; (6) grew and reproduced by the same type of simple cell division as living cyanobacteria and other microbes, shown by preserved partially divided cells; and (7) are present in the rock together with coaly organic debris that by its chemical makeup (its carbon isotopic composition) is certain to have been produced by photosynthesis, as we will see in chapter 6.

9. What kind of organisms are they, and why do they matter?

All of the Apex fossils are remnants of cellular filamentous microbes known as prokaryotes, an early-evolving type of microorganism in which the hereditary material (DNA) is in simple strands within the cell rather than packaged in a cell nucleus as in more advanced forms of life (eukaryotes). Among the prokaryotes, all of the Apex fossils belong to the domain known as the Bacteria (which includes cyanobacteria as well as less advanced bacterial types) rather than the Archaea, the other prokaryotic domain (a rather recently discovered branch of the Tree of Life made up of microbes that often live in harsh, high-temperature, acidic settings).

But precise placement of the Apex fossils among the Bacteria is less certain. My view is that six of the eleven species probably are cyanobacteria; three, members of more primitive bacterial stocks; and two could be either cyanobacteria or bacteria. These assignments are consistent with the size, shape, and organization of the cells in the fossils; the way they divided as they grew; the way they broke down as the microbes were preserved (their "taphonomy"); the geology of the site and the mineral makeup of the associated sediments; the carbon isotopes of the Apex organic matter; and cell-by-cell comparison of the Apex fossils to members of younger Precambrian communities and living filament-forming prokaryotes, both cyanobacteria and bac-

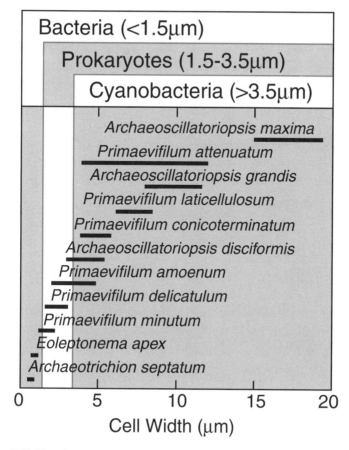

Figure 3.13 The eleven named species of Apex microorganisms, ordered according to the width of their body cells, showing that the three smallest (less than 1.5 μm wide) are classified as "probable bacteria"; the two next smallest (1.5 to 3.5 μm broad) as either bacteria or cyanobacteria ("undifferentiated prokaryotes"); and the six largest species as "probable cyanobacteria."

teria. In fact, several of the larger Apex species so closely resemble modern and Proterozoic cyanobacteria that it seems to me likely they are members of a common and particularly well-known subgroup that still lives today (a formal taxonomic family of cyanobacteria, the Oscillatoriaceae).

If I am right about these relations, the presence of cyanobacteria in this nearly 3,500-Ma-old community tells us that early evolution proceeded very far very fast. All cyanobacteria are able to do the kind of photosynthesis that gives off oxygen; and, like higher plants and animals, all can breathe in oxygen (by the process known as aerobic respiration). Both of these, however, are advanced ways to live, evolved from more primitive ways in which free oxygen plays no part. So, if cyanobacteria existed at this early time, the earlier evolved processes must also have been present; the living world would have to have included organisms that photosynthesized without giving off oxygen (bacterial photosynthesizers) as well as those that produced it (cyanobacteria), and microbes that lived in the absence of oxygen (anaerobes) as well as those that breathed it (aerobes). These are precisely the same processes that power the present-day living world. If cyanobacteria are represented among the Apex fossils, we are forced to conclude that the basics of the world's ecosystem had already become established by this very early stage in the history of the Earth.

No one has publicly disagreed with my interpretation of the Apex fossils. But, privately, some would prefer I were mistaken, since they (and I, too) would prefer a simpler evolutionary story, one that told us these oldest fossil organisms were capable only of primitive ways of living and that advanced metabolic lifestyles evolved much later. But the evidence seems strong, and what one might "prefer" shouldn't matter. As Darwin, Cloud, and others taught us years ago, there is only one court of last resort—the fossil record!

Though much is yet unknown about the Apex community, it seems certain that it differed from stromatolitic communities of the Proterozoic. In these younger deposits, stromatolite-building filaments tend to be splayed out over the surface of a stromatolite, densely intertwined in stacked, matlike layers. This is not true of the Apex microbes. Instead, they are arrayed at random and preserved usually as solitary individuals rather than interwoven in carpetlike sheets. And unlike Proterozoic filaments, like raisins scattered through a loaf of raisin bread they "float" in thick wispy masses of what originally was gelatinous mucilage. Many prokaryotes and almost all cyanobacteria secrete mucilage from their cells, but the Apex community is the only one known where the microorganisms lived embedded in such massive clumps.

Because this is the only microbial community known from rocks so ancient, it is not possible to say whether secretion of copious mucilage was typical. But it could be that this sticky secreted mass glued the Apex organisms to the shallow seafloor, where it enabled them to harvest sunlight protected from damaging UV rays by overlying seawater. If so, like other traits characteristic especially of cyanobacteria (their ability to glide away from intense light and to manufacture biochemicals both to protect themselves from harmful UV and to repair the damage it causes), the extracellular glue played a role in helping these early evolving microorganisms cope with an inhospitably harsh environment.

10. Will older fossils ever be found?

We can only guess about this and there is no guarantee we'll be right. Years ago, when the oldest rocks known dated from only 3,100 Ma ago, I asked a world-famous geochronologist whether more ancient units would ever be found. He assured me there was "no way this could happen" because the world's oldest terrains had been "exhaustively searched," yet reports soon surfaced of 3,750-Ma-old rocks in southwestern Greenland.

I don't know how far back in time the fossil record will eventually be traced. Yet because the search has barely begun, it seems likely that fossils older than the Apex filaments will turn up, perhaps in the Western Australian Pilbara, still the most promising region known. But I am doubtful that the rock record will ever yield evidence of the origin of life itself. The very first forms of life can hardly be expected to have built stromatolites or carried out complicated biochemical processes of the type that leaves an isotopic signature. They probably didn't even have fossilizable cell walls, as we will see in chapter 5, and were made of chemicals far too fragile to be geologically preserved. One can certainly hope that the future holds new ways to research this problem, but the prospects do not seem promising.

The Oldest Fossils Known

Though only the first few halting steps have been taken in the search for the earliest records of life, progress is encouraging, spurred by development of five rules to sort the bona fide from the bogus: (1) The source of the fossil-bearing rock must be established beyond doubt, normally by replicate sampling, and (2) its age must be well documented, ideally by high-resolution U-Pb zircon dating. Studies of thin sections must demonstrate that the fossil-like objects (3) are part of the rock, rather than contaminants, (4) entombed when the rock was formed. And it must be shown that the objects (5) are unquestionably biologic rather than "fossil-like" minerals or other pseudofossils.

The oldest fossils meeting these tests are eleven kinds of cellular threadlike microbes petrified in the 3,465-Ma-old Apex chert of northwestern Western Australia. These are recent finds on which further study is needed. But it already seems clear that all are prokaryotes of the Bacterial domain and the best evidence is that the fossils include several types of oxygen-producing and oxygen-breathing cyanobacteria. If so, these organisms are not only extremely ancient but surprisingly advanced, and show that early evolution proceeded faster and farther than anyone imagined.

How Did Life Begin?

The Basics of Biology

One of the most pleasing things about science is that there are no dumb questions. Nobel Prize winners often wonder about the same things the rest of us do. And the universal question we all seem to ponder is, how did life begin? Before we tackle that great puzzle we need to be clear about a few terms that will help us understand what life is.

What Is a Cell?

Everything alive is made of cells, compartments that wall off the living juices of an organism from its surroundings. Cells house the biochemical basics of life: gene-containing chromosomes, enzymatic proteins, and the machinery of metabolism. In the smallest forms of life—parasitic mycoplasma (some only 0.1 μm across, one ten-millionth of a meter)—the separating boundary is a thin, flexible membrane. In larger cells, such as bacteria, these combine with a thick cell wall to form a tough outer casing. Early-evolving organisms were single celled, sheltered in tiny one-roomed houses. Colonies of clustered cells and many-roomed multicellular forms of life evolved later.

What Is a Gene?

A gene is the basic unit of heredity. It consists of a chainlike sequence of hundreds of linked chemical units called nucleotides, each composed of three subunits: a sugar, a phosphate, and a nitrogen-containing base, either a purine (adenine, "A," or guanine, "G") or a pyrimidine (cytosine, "C," or thymine, "T"). The sugar of each nucleotide is attached to the phosphate of its nearest neighbor, linking together one

strand of an exceedingly long molecule known as DNA (deoxyribonucleic acid). Each DNA molecule consists of two such strands,
wound in a double helix and made up of many genes aligned one after
the other like boxcars of a very long train. The gene-bearing DNA is
housed in chromosomes: twenty-three pairs of banded, rodlike structures in the nuclei of human cells, a single threadlike loop in nonnucleated microbes such as bacteria. The sequence of nucleotides in
each gene, the order of the A, G, C, and T bases, serves as a chemical
code instructing the manufacture of a protein at one of the many thousand minute protein factories called ribosomes scattered throughout a
cell.

What Is a Mutation?

A mutation is any change in the normal sequence of nucleotides in a
gene, such as the deletion of a nucleotide or the switch of one nucleotide to another. Mutations result most often from mistakes made
when the instructions encoded in a gene are copied, to be forwarded
to a ribosome, but can be caused also by radiation, by heat, by chemical mutagens (such as the cancer-causing agents of asbestos, cigarette
smoke, industrial pollution), even by viruses. Human chromosomes
are made up of 80,000 genes and about 3 billion nucleotides, so there
are plenty of opportunities for mutations to arise. The instructions
coded in a mutated gene have been changed, and when these arrive at
a ribosome an altered protein is usually produced, one almost always
less able or unable to function normally. Cells like those of humans
with two copies of each gene can often get by with one healthy version. But a mutation can be deadly if it occurs in an organism with
only a single copy of its genes, like many primitive forms of life, so
mechanisms to fix damaged genes and their protein products evolved
early in biologic history.

What Is an Enzyme?

Enzymes are chemicals, usually proteins, that hasten biochemical processes by acting as matchmakers that bring together molecules and
speed their interaction. Life depends on the steady presence of a complicated, often interdependent sequence of chemical events such as
the breakdown of a molecule, the transfer of one part of a molecule to

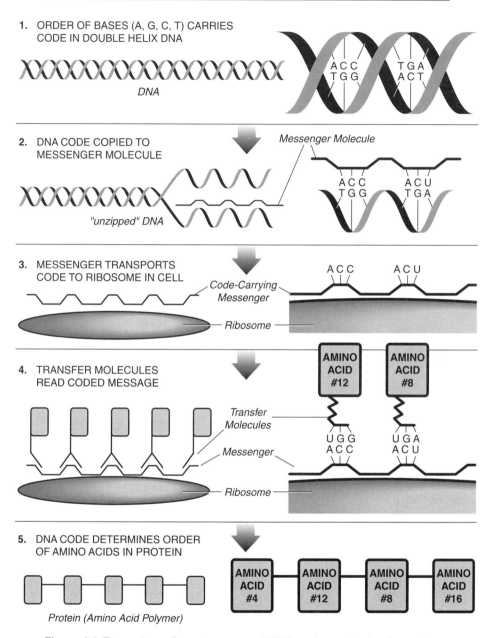

1. ORDER OF BASES (A, G, C, T) CARRIES CODE IN DOUBLE HELIX DNA

DNA

ACC TGG   TGA ACT

2. DNA CODE COPIED TO MESSENGER MOLECULE

"unzipped" DNA

Messenger Molecule

ACC TGG   ACU TGA

3. MESSENGER TRANSPORTS CODE TO RIBOSOME IN CELL

Code-Carrying Messenger

Ribosome

ACC   ACU

4. TRANSFER MOLECULES READ CODED MESSAGE

Transfer Molecules

Messenger

Ribosome

AMINO ACID #12   AMINO ACID #8

UGG ACC   UGA ACU

5. DNA CODE DETERMINES ORDER OF AMINO ACIDS IN PROTEIN

Protein (Amino Acid Polymer)

AMINO ACID #4 — AMINO ACID #12 — AMINO ACID #8 — AMINO ACID #16

Figure 4.1 The pathway from chromosomal DNA to the synthesis of proteins at ribosomes.

another, or the assembly of a molecule from smaller pieces. Even the simple act of breathing—taking in oxygen and using it to burn foodstuffs to generate energy—involves many steps, each controlled by a different enzyme. As in breathing, each event in every biochemical process is governed by a different enzyme. Life requires so many enzymes that most of the chromosomal genes are earmarked for their production.

What Is Metabolism?

The energy to power cells comes from the enzyme-speeded break-down of foodstuffs. These can either be manufactured within a cell (for example, by photosynthesis, as in plants, cyanobacteria, and pho-tosynthetic bacteria) or be taken in from the surroundings (as in ani-mals, fungi, and various kinds of microbes). "Metabolism" refers to both the buildup of foodstuffs and their energy-generating breakdown. Early-evolved metabolic processes (such as the manufacture of food-stuffs by photosynthetic bacteria or its breakdown by fermentation) are anaerobic, happening in the absence of free oxygen, whereas more advanced metabolism gives oxygen off (as in the photosynthesis of plants and cyanobacteria) or consumes it (as in aerobic respiration, "breathing").

What Is the Tree of Life?

As we saw in chapter 2, the worldwide search for fossil evidence of ancient life began in earnest in the mid-1960s. Only a decade or so later, a way emerged to test the completeness of this new-found fossil record by using the molecular makeup of organisms to construct a Universal Tree of Life showing the evolutionary relations among *all* organisms living today. This new approach is possible:

1. Every organism manufactures proteins (such as the enzymes that control metabolism) at tiny globular bodies known as ribosomes.

2. These protein factories are made mostly of ribonucleic acids (RNAs), long single-stranded molecules that, like DNA, have a sugar-phosphate backbone from which extends a riblike array of four nitrogenous bases, A, G, C, and U (uracil, in place of the thy-mine of DNA).

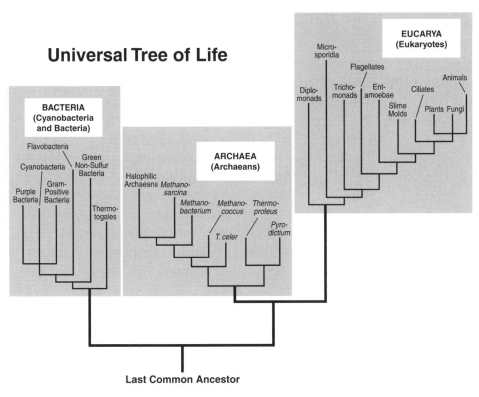

Figure 4.2 The rRNA Universal Tree of Life.

3. Mutations in an organism's DNA can alter the sequence of bases in ribosomal ribonucleic acid (abbreviated "rRNA").

4. Because members of groups closely related by evolution have similar DNAs, their rRNAs are also similar, while more distantly related organisms have more dissimilar rRNAs.

Comparison of the sequences of nitrogenous bases in the rRNAs in living members of the major groups of life gives a clear index of how closely or distantly related the organisms in them are. The resulting Universal Tree of Life (based on a kind of rRNA known as "16S"), constructed so that the vertical length of each branch corresponds to the number of base sequence changes that occurred since the branch split off from its nearest neighbor, shows four main facts:

1. Though plants and animals are the most familiar kinds of life, they are merely two of more than twenty major evolutionary branches.

2. The Tree of Life is overwhelmingly composed of microscopic organisms; of the more than twenty branches only three (plants, fungi, and animals) contain forms large enough to see without a microscope, and each of these contains microsopic forms as well.

3. Every present-day organism belongs to one of but three major groups (known formally as domains): (1) Eucarya, eukaryotes (having cells in which chromosomes are packaged in a saclike nucleus); (2) Archaea, archaeans (nonnucleated microbes including the only organisms that give off methane and many "extremeophiles" that thrive in exceedingly acidic, high-temperature settings); and (3) Bacteria, the domain that includes oxygen-producing cyanobacteria and all of the many different kinds of bacteria.

4. Eukaryotes are more closely related to archaeans than to bacteria, and the last common ancestor of all present-day life—the root of the Universal Tree of Life—lies between the Bacteria and the Archaea.

The branching pattern of the Tree of Life agrees fairly well with the known fossil record—Bacteria and Archaea early, Eucarya much later. But rRNA trees cannot show precisely when the various branches sprouted. Such trees are based on organisms alive today, so if they clocked evolution accurately all branchtips would extend to the same height, corresponding to the present, and the top of the tree would be flat like a tall Christmas tree with its top shaved off to fit a low room. But it is not. It zigzags up and down because its branches evolved at different rates. The long-branched groups (such as Eucaryal microsporidia and diplomonads) evidently evolved faster, whereas groups with shorter branches (several in the Archaea and eukaryotic ciliates, plants, and fungi) evolved more slowly.

Accurate dating of life's early branching remains a challenge. There are good reasons to hope that the molecular makeup of living cells holds the key, but neither rRNA trees nor dates based on evolving proteins has yet proved reliable. Likewise, early fossils are still too incompletely known to provide precise answers and, in any event, can record only the first *detected* occurrence of a biologic group, not its first *actual*

ELEMENTAL COMPOSITION					COMPONENT OF LIFE	COMPOSITION OF REPRESENTATIVE ORGANISMS								
Carbon	Hydrogen	Oxygen	Nitrogen	Phosphorus / Sulfur		Lettuce	Celery	Mushroom	Oyster	Codfish	Bacterium	Cow	Chicken	Pig
H		**O**			WATER (Universal Solvent)	95%	94%	90%	88%	83%	75%	74%	66%	57%
C	**H**	**O**	**N**	**S**	PROTEIN (Enzyme Catalysts)	1.3	1.4	3.6	6.0	12.0	17.5	19.6	21.2	20.1
C	**H**	**O**			FAT (Energy Storage)	0.4	0.4	0.4	1.5	3.5	2.5	4.2	11.0	20.2
C	**H**	**O**			CARBOHYDRATE (Cell Walls)	2.1	3.0	5.1	2.4	0	1.3	0	0	0
C	**H**	**O**	**N**	**P**	DNA, RNA, ATP (Genes, Energy)	1.2	1.2	0.9	2.1	1.5	3.7	2.2	1.8	2.7

Figure 4.3 In addition to CHON and P and S, numerous other chemical elements (magnesium, iron, copper, and cobalt, for example) are also required by living systems, but only in trace amounts.

presence. Even abundant paleobiologic evidence can never do more than provide minimum ages for branches of the Tree of Life.

The Universals of Life

Living systems are less complicated than one might think. All are made of a surprisingly small number of chemical ingredients. You, I, and the rest of the living world—even a head of lettuce—are made mostly of water. This watery makeup is a heritage of the distant past. Because life originated in watery surroundings almost all its chemistry is carried out in water, the liquid medium that makes up the sap, the cytosol, of living cells.

Life consists mostly of four chemical elements: carbon, hydrogen, oxygen, and nitrogen—CHON—together sometimes with S (sulfur) and P (phosphorus). These four prime elements of life make up more than 99.9% of all living systems. Why CHON? Why not titanium, gold, krypton, and thulium, or some other exotic mixture? The answer is simple. Life is made of CHON because these elements are plentiful, four of the five most common in the Universe. (The fifth abundant element, helium, doesn't count. Helium is inert, nonreactive—a great gas for balloons, but an element unable to join with others to form robust chemical compounds.) There was plenty of CHON around when life got started. Moreover, all these elements are able to combine with one another to form small sturdy molecules such as methane (CH_4), carbon dioxide (CO_2), and ammonia (NH_3), compounds that because they dissolve in water (H_2O, another linked pair of the prime elements) can play an active role in the workings of life.

From protists to petunias, microbes to man, living systems are wondrously diverse. Yet at a chemical level, all are practically the same, a sameness showing that all life stems from a common stock. All life is made of CHON(SP) and is composed of the same three dozen kinds of fundamental CHON(SP)-containing building-block molecules. These small compounds, such as amino acids, sugars, and the purines and pyrimidines of DNA and RNA, are called monomers, and in all are linked together to form the same few kinds of large polymer molecules (such as proteins, carbohydrates, and nucleic acids) so important to life. The sameness carries over to metabolism as well, since energy to power life is produced in only a few closely related, more or less similar ways.

These universals show that not just the millions of species living today but *all organisms through all of time over the entire planet* trace their roots to a single primal cell line.

To understand how this cell line nestled at the root tip of the Tree of Life came to be, we need to answer three crucial questions. (1) How did CHON-containing monomers arise on the lifeless early Earth? (2) How did these small compounds link together into large polymers like enzymatic proteins and hereditary DNA? (3) How did cells and the life-sustaining machinery of metabolism arise? Answers to the first two questions are fairly well understood. The third, as we will see in chapter 5, is more enigmatic.

How Did Mononers of CHON Arise on the Lifeless Earth?

Darwin's Deduction: Monomers of CHON *Might* Be Easy to Make

The first of the crucial questions has been studied longest. Indeed, the notion that CHON-containing monomers might be easy to make on the primordial Earth dates from Darwin and a famous passage in a letter he wrote in 1871 to his botanist friend, Joseph Hooker:

It is often said that all of the conditions for the first production of a living organism are now present, which could ever have been present. But if (and oh! what a big if!) we could conceive in some warm pond, with all sorts of ammonia and phosphoric salts, lights, heat, electricity etc. present, that a protein compound was chemically formed ready to undergo still more complex changes, at the present day such matter would be instantly devoured or absorbed, which would not have been the case before living creatures were formed.

The life-starting scenario dreamed by Darwin is astonishingly close to what we think today. But only the rudiments of biochemistry were known in Darwin's time, and he was a naturalist interested in the riddle of life's evolution, not its origin. The problem of life's beginnings remained in limbo until the 1920s when a solution finally came into focus, influenced—through a fascinating if unlikely chain of events—by Darwin's own teachings.

The History of a Hypothesis: From Darwin to Oparin to Us

Throughout human history, the origin of life has posed one of the most profound mysteries in all of nature. The major breakthrough came in 1924 when Aleksandr Ivanovich Oparin (1894–1980), a young Russian plant biochemist, authored a small book outlining how CHON-containing molecules might have formed before life began and given rise to the first cells. Three decades later Oparin's notion bore fruit when Stanley Miller, a second-year graduate student at the University of Chicago, showed in a masterful set of experiments that amino acids are indeed made spontaneously under conditions similar to those of the ancient lifeless planet.

Because Oparin was the first to suggest a plausible solution to the origin of life, he was lionized. At an early age he was elected to the prestigious USSR Academy of Sciences; his many books were translated worldwide; and he received accolades and medals, both foreign and domestic, including the coveted title Hero of Socialist Labor. He was treated almost like a head of state.

Academician Oparin was a big man, not only in reputation but in height, girth, and carriage. I first met him on one of my early trips to

Figure 4.4 Academician A. I. Oparin and Bill Schopf at UCLA (November 1976).

the USSR in 1972, and over the years we developed a strong friendship. I have visited many times the A. N. Bach Institute of Biochemistry that he headed in Moscow, and was recently honored to be the first foreign member elected to its Scientific Presidium.

In the autumn of 1976, I helped arrange for Oparin to spend two months as a senior visiting professor in my UCLA laboratory. At the time this was somewhat unusual—the Cold War still continued, at least in the minds of some of my colleagues who let me know they were not pleased with the prospect of rubbing shoulders with a Russian communist. But Oparin came as a scientist, not an apparatchik, and his visit was not only an opportunity for my students to get to know the great man but for Oparin to be exposed to the students' views of Western values and ideas (democracy, capitalism, individualism, a free press) about which he was surprisingly naive. Each week, a dozen or so students and I had lunch with Oparin and his ebullient wife, Nina Petrovna. My Russian teacher graciously agreed to translate the conversations. At one such get-together, Academician

Figure 4.5 A. I. Oparin with UCLA students, Bill Schopf's Russian teacher, and Schopf.

Oparin told us how he had first come to be interested in the origin of life, a remarkable story that reveals much about the human side of science.

During his youth in Uglitch, a rural center of wooden houses, farmers' markets, gas lights, horse-drawn carts, and muddy roads situated north of Moscow on the Volga River, Oparin developed a keen interest in botany. He amassed a collection of local plants and even conducted simple experiments on plant growth. In 1912, as a winner in national examinations, he was awarded the opportunity to enroll at Moscow State University, even today the most renowned of all Russian institutions of higher learning. That spring his high school science teacher arranged for him to visit the university. But on the appointed day they missed the morning train and arrived late, in time to attend only the last lecture of the day.

Given the opportunity to sit in on but a single class, Oparin chose that of the foremost Russian botanist of the day, K. A. Timiryazev (1843–1920). His choice came as no surprise to his high school teacher, for Oparin had spoken often of the famous botanist, the author of the classic text *The Life of Plants*, which Oparin had learned by heart and a scientist Oparin would come to hold in the highest esteem as his "first teacher." Moreover, though almost 70 years old, Professor Timiryazev was a highly regarded popularizer of science, widely known for his enthusiastic public lectures, and moreover, he was a confirmed Darwinian, a point of view rarely espoused in czarist Russia and one that much intrigued the young Oparin.

Timiryazev spent the lecture telling how he had become a proponent of Darwin's views. Little more than a decade after publication of *On the Origin of Species*, Timiryazev, then a recent graduate of Moscow State University, traveled to Down House—Darwin's 20-acre estate in Kent, outside London, nestled among the gently sloping hills, "downs," for which it is named—hoping to meet the great naturalist. Darwin was ill, as he often was in middle and later life, and was not receiving visitors. But Timiryazev would not be put off. He rented an upstairs room at the local pub down Luxted Road and returned daily for more than a week, sitting patiently on the front stoop, waiting for his opportunity. Finally, Darwin granted the young Russian scholar an audience. They strolled the "sandwalk" behind Down House, the pebbled track through the woods where Darwin often went to think, and talked of evolution. By the end of their gentle walk, Timiryazev was convinced: Darwin *must* be right!

The young Oparin was enthralled, and soon he, too, was convinced. But as he listened to Professor Timiryazev, he spotted what seemed a gaping hole in the theory. Darwin had dealt magnificently with the evolution of animals and Timiryazev with the evolution of plants. But where did animals and plants come from? "Darwin had written the book, but it was missing its very first chapter," Oparin told the students that day at lunch, and they were spellbound as he told them how he had devoted his long career to filling in the pages of this missing first chapter of life's history.

My recollections of that luncheon are vivid. I realized that thanks to Oparin, the students and I could now count ourselves among Darwin's true academic descendants, tied through the generations by an amazingly short skein of human interactions—from Darwin as an old man to Timiryazev in his youth, from an aging Timiryazev to the aspiring young Oparin, from the elderly Oparin to those around that lunch table. It was thrilling!

REALIST MEETS SURREALIST

A year or so after I first met Academician Oparin, we came together again at the June 1973 triennial meeting of the International Society for the Study of the Origin of Life (ISSOL). Held in Barcelona, Spain, the conference was a week-long affair attended by several hundred scientists from over forty countries and hosted by Juan Oró, the foremost Spanish worker in the origin of life field. After the conference, Oparin was to journey up the Catalan coast to visit the surrealist artist Salvador Dalí, a personal friend of Oró who had prepared special artwork for the conference booklet. It was my good fortune to be invited to tag along.

Oparin, of course, was famous. But Dalí was even more so, renowned worldwide not only for his talent but his eccentric behavior—such as when he delivered a public lecture in London clad in a deep-sea diving suit and helmet, or when he hiked up the middle of New York's Fifth Avenue breathing through a snorkel as he peered out of a goldfish bowl strapped about his head, fish swimming before his eyes. The Oparin-Dalí meeting was certain to be memorable!

After spending the night at one of the opulent resorts on the Costa Brava north of Barcelona, we set out over dusty roads in a new VW-microbus to find Dalí's home at the village of Gerona. At first sight, his house was not in the least impressive. Situated at the edge of a small inlet, like other houses of the fishing village it was a smallish, unremarkable, whitewashed stucco structure with protruding dark wood beams. It was so normal that we wondered whether we had found the right place. But we had, the tip-off given by an old weathered rowboat—with a two-meter-high sapling rising from its floorboards—beached in his front yard.

We were greeted at the door by a coiffured, uniformed, rather matronly maid who told us we were expected and that Señor Dalí and his wife

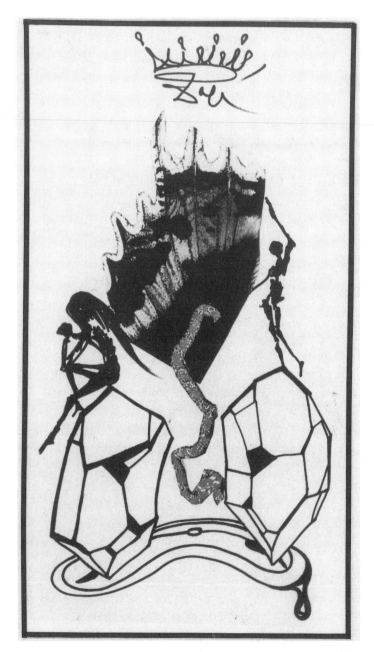

Box fig. 4–1 Dalí's painting on the cover of the program of the 1973 ISSOL meeting in Barcelona (used again when ISSOL met in Barcelona in 1993). Dali's signature and crown logo are at the top.

awaited us in the garden. We were then ushered into a small dimly lit entryway—a semi-alcove made even smaller by the presence directly in our path of a stuffed full-grown brown bear that, with mouth agape and arms outstretched, glowered from its pedestal. This was not a run-of-the-mill museum display: the bear was decked out in a top hat and cape. Over one arm was draped an umbrella, the other paw held a mirror-and-jewel-encrusted walking stick, and pinned on the bear's chest and dangling from colored ribbons around its neck were more than a dozen medals and medallions. Apparently the bear served as a handy repository for baubles bestowed on the señor over the years.

The bear was quite a sight, certainly worth a picture! But as I unslung my camera and began to afix the flash attachment, I was informed by the maid—politely but firmly—that one does not take pictures in Señor Dalí's home. Should such a need arise the señor's personal photographer would be available. I was disappointed. But I was of course agreeable to abide by the house rules.

Our little group started off along a corridor leading toward the garden. The hallway turned out to be long, circuitous, and surprisingly narrow. It linked what had once been three separate fishermen's homes, and Dalí later told us he had fashioned this tortuous confine to remind him of his unpleasant stint in prison in 1924 for political activism (but according to his detractors, who note that Dalí was incarcerated for little more than a week, the passage was actually designed this way, then unveiled to the press in an effort to repair Dalí's tarnished image as an admirer both of Franco and Hitler).

Oparin's girth was rather too large for the passageway so he and his wife, Nina Petrovna, turned back to find an alternate route as the rest of us tunneled on. We finally emerged into a small room—paintings on one wall, in a corner hanging from the ceiling a child's medieval suit of armor encapsulating a meter-high inflated plastic astronaut—and then entered onto the garden. There before us were Dalí, his wife Gala (born Elena Diaranov, a Russian emigrant and Dalí's muse and business manager), his official photographer, a couple of maids, and a few guests. The Oparins soon arrived and we settled down to canapés, pink champagne, and innumerable toasts.

Dalí was a surprise. By no means as theatrical, even maniacal, as I had expected, in social conversation he was inquisitive, thoughtful, actu-

Box fig. 4–2 Salvador Dalí, Aleksandr Ivanovich Oparin, and Oparin's wife, Nina Petrovna.

ally rather shy. He and Oparin got along famously, each obviously curious about the other, and despite a language barrier (Oparin speaking Russian only) they carried out a lengthy animated conversation—partly via body language, each pointing, nodding, and gesturing broadly, and partly through Nina Petrovna, a former English teacher at Moscow State University and an experienced translator.

The small secluded garden was festooned with Dalí originals: a pink satin love seat fashioned after the famed lips of Mae West (the "why-don'tcha-come-up-and-see-me-sometime" actress) and surrounded by gauze-draped cardboard Pirelli tires; a full-size papier-mâché camel, on one flank lettered with the slogan "Camel Filters!," situated in the midst of a dilapidated cactus garden; an elongate fishless reflecting pool encompassed by a well-worn lawn of coarse green plastic grass and fronted by a huge lighthouse lamp framed in a Persian cupola. Soon Dalí put over the loudspeakers a tape of the Kremlin Tower bells he had made one February in Moscow. On a sunny summer afternoon in northern Spain, immersed in products of Dalí's bizarre imagination, we sipped champagne and listened to the cacophony of Red Square in winter.

Box fig. 4–3 Satin love seat modeled after Mae West's lips.

Box fig. 4–4 Cactus garden papier-mâché camel.

Box fig. 4–5 Fishless reflecting pool with Señor Dalí in the right foreground, his back to the camera.

Box fig. 4–6 Lighthouse lamp and cupola.

After he and Dalí had talked for awhile, Oparin decided that he wanted a photograph of the two of them in conversation. Through Nina Petrovna, Oparin proposed this to Dalí, but the official photographer was nowhere to be seen so Dalí suggested they wait for a moment. Oparin evidently had no idea what Dalí had said so he gestured to me, raised his hands as though he had a camera in them, and bellowed something in Russian ("Beeel . . . "). I didn't understand his words, but I did understand the gestures and began taking pictures as fast as I could for fear Dalí's pho-

Box fig. 4–7 Salvador Dalí and Aleksandr Ivanovich Oparin (an "unofficial" photo taken at Oparin's request).

tographer would return to take my place. (After awhile he did, but with my camera now unloosened I managed to get some candid photos of the surrealist and his realist guest, though I still regret missing the photo-op of the bemedaled bear!)

Oparin's Opinion: Monomers of CHON
Should Be Easy to Make

While a graduate student at Moscow State University, Oparin became absorbed by the problem of life's origin. At the time, many scientists thought the first living systems were probably simple algae, tiny single-celled phytoplankton that arose spontaneously (by an unknown, but presumably quite improbable string of events) in an environment more or less the same as now. Algae were pegged as the first organisms because they were considered the simplest photosynthesizing forms of life, able to grow using easily available light, water, and carbon dioxide, and all-important as the base of the food-chain without which an ecosystem of eaters and eatees could never become established.

Oparin had a different notion. Schooled in botany, he was impressed by the complex inner workings of plant cells, which he viewed as more elaborate than those of animals. Animals and animal-like organisms (protozoans and fungi, for example) are eaters, sustaining themselves by taking in food and breaking it down to generate energy and the molecular building blocks to construct their bodies. In contrast, plants and plantlike organisms (cyanobacteria and single-celled algae, for example) live by a more complicated two-step process: first, they carry out photosynthesis, capturing light energy and storing it in the organic molecules they build up; second, they break down these molecules—the same way animals do—to provide energy and the molecular ingredients from which to build their cells. Oparin reasoned that if life's evolution followed the path from simple to complex, animal-like eaters (technically, "heterotrophs") would have come first and more complicated plantlike eatees (autotrophs) only later. This novel notion of a heterotrophic origin of life flew in the face of prevailing scientific sentiment.

Though Oparin's vision fit with a Darwinian simple-to-complex view of evolution, it was at odds with the necessities of the food chain. If the first forms of life were animal-like, they would have had no photosynthesizers to feed on—what did they eat? Again, Oparin's botanical knowledge came to the fore. Most scientists assumed that life began while the atmosphere was essentially like it is today. But Oparin knew that the oxygen of the present-day atmosphere is a by-product of photosynthesis, so if plantlike organisms evolved later than heterotrophs, there initially would have been no atmospheric oxygen. Moreover, he realized that free oxygen combines readily with organic substances in the chemical process of oxidation, burning. So he reasoned that if the early environment lacked free oxygen, then simple organic compounds, formed by the action of volcanic heat or lightning on CHON-containing atmospheric gases, would accumulate rather than be burned and destroyed, and these would have dissolved in the early seas to form an organic-rich broth. This postulated primordial soup was the linchpin to Oparin's scenario—over time, he thought, some of its ingredients could have linked together to give rise to life while others served as fodder for the budding life-forms.

Oparin's concept is straightforward. The environment of the lifeless Earth lacked free oxygen, so chemical processses could spontaneously give rise to the organic constituents of the first cells, simple animal-like heterotrophs that fed on the primordial soup from which they had emerged. Life evolved from simple to complex, not the other way around, and the base of the primal food chain was provided by *non*-biologic chemical reactions, not single-celled phytoplankton.

Oparin wrote up his views, and in 1918, while still a graduate student, submitted a manuscript for publication. This was less than a year after the Bolsheviks seized power, and though much of the czar-ist regime had been cast aside, its official censors and their fiercely enforced rules had not. Prior to the revolution the czar had reigned supreme, not only over the Russian state but over the Russian Ortho-dox Church as well. And to Russian Orthodoxy, the origin of life was sacrosanct. Oparin's manuscript was rejected out of hand.

Oparin later claimed this rebuff was a godsend, for despite his confident demeanor and genteel bow ties he was at heart an inex-perienced 24-year-old Uglitch farm boy. Prodded by the rejection, he worked and reworked his manuscript over the next four years, coming to realize that his first effort had missed the mark. In 1922 he pre-sented a formal report on his novel notion (at a meeting of the local branch of the All-Russia Botanical Society) and submitted an up-graded version of his opus for publication. This first volume, this "little pamphlet" (as he affectionately called it) on *The Origin of Life*, published in 1924 and only seventy-one pages long, was the most important of the dozens of books Oparin wrote. Even today it has enormous influence on understanding how life started.

Miller's Milestone: Monomers of CHON *Are* Easy to Make

Oparin has not always been honored for his breakthrough idea. The world was different in the 1920s—no jumbo jets, fax, E-mail, or international CNN—and few scientists outside the Soviet Union were capable of reading Russian. (And Oparin, even in his later years, knew only a single phrase in English, which he delivered with great gusto at the conclusion of each translated lecture, a guttural, booming "Zankyuverymutgh!"). Unsurprisingly, the 1924 pamphlet (like the 1922 presentation) went unnoticed internationally.

It was not until 1938, when his first major book was translated into English, that Oparin's views became more widely known. Their reception, however, was lukewarm. His scenario fit well with Darwinian evolution, but it also meshed with Marxist dialectic materialism (since it assumes a material reality that changes from one stage to the next in a process governed by the inherent properties of matter). To some this Marxist fit was disturbingly "too convenient," suggesting Oparin's views were more political than scientific, a skepticism bolstered by Oparin's friendship with the Stalinist geneticist T. D. Lysenko (whose summer dacha adjoined Oparin's on the outskirts of Moscow), and the similarity of Oparin's thesis to that proposed in the late 1920s by J. D. Bernal, a famed British biologist well known for his leftist leanings.

A second, even more telling hurdle was that Oparin's scenario was purely theoretical—a plausible chain of ideas devoid of experimental backing. Theory first, experiment second may be the gold standard in quantum physics, but in the life sciences, observation usually leads the way and theories are only as firm as the real-world facts on which they stand. Here, Oparin failed. To him, his theory was sound enough to stand alone. But it never would have caught on were it not for the experimental support provided by Stanley Miller decades later.

In September 1951, Stanley Lloyd Miller, a newly minted graduate of the University of California, Berkeley, enrolled in the doctoral program of the Department of Chemistry at the University of Chicago. During his first semester he attended a departmental lecture on the origin of the solar system in which Professor Harold C. Urey noted— almost in passing—that the Earth's primordial atmosphere would have been favorable for formation of simple organic molecules, precursors of the first living cells.

Miller was intrigued. Urey, a Nobel laureate (awarded in 1934 for discovery of deuterium, an isotopic form of hydrogen) and the world's expert on the early solar system, was unlikely to be wrong. His argument made good sense: (1) Because hydrogen is the most abundant element in the solar system, the Earth must have started out hydrogen-rich. (2) Gases enveloping the forming Earth would there-

Figure 4.6 Stanley L. Miller at the triennial meeting of the International Society for the Study of the Origin of Life in Barcelona, 1973.

fore be hydrogenated (technically, "chemically reduced") and the primordial atmosphere made of molecular hydrogen (H_2), methane (CH_4), ammonia (NH_3), and water vapor (H_2O). (3) In this setting, chemical reactions powered by lightning or high-energy UV light would produce hydrogen-rich organic molecules like those that make up living systems, so this nonbiologic way of forming organic substances was a promising path to the origin of life.

Urey's claim was that in a hydrogen-rich, oxygen-free primitive atmosphere it would be easy to form reduced biologiclike molecules because relatively few chemical bonds in the hydrogenated gases would need to be broken and put together in new configurations. Neither Miller nor Urey was aware of it, but this fit well with Oparin's idea of a prebiologic primordial soup.

About a year later it was time for Miller to select a suitable topic for his doctoral research. He had thought he might like to work with Professor Edward Teller and study the genesis of chemical elements in stars, but that fell through when Teller formally left Chicago to head the super-secret H-bomb project at Los Alamos National Laboratory, New Mexico. In September 1952 Miller sought out Professor Urey and explained that he wanted to do a laboratory experiment testing Urey's idea of synthesizing biologiclike organic monomers in a reducing atmosphere.

Urey was pleased, but not overwhelmed. His interests had shifted and he was now trying to understand the detailed nature of the embryonic solar system. A crucial clue, he thought, would come from knowing the abundance of the element thallium in various kinds of meteorites. He proposed this project to Miller. But as Miller later recalled, "Thallium was very nice, but I was more interested in organic compound synthesis and I hinted that it was either organic compound synthesis or I would go to another professor for a thesis project." Urey relented, but because the project was "unusual . . . and the chances for success quite small," he gave Miller "six months or a year" to succeed; otherwise, Miller's doctoral research would be switched to some less chancy subject like the thallium project.

Given the go-ahead, Miller went to work full steam. Over the next few weeks he mulled through Oparin's 1938 book and Urey's most recent (1952) paper on solar system origins, and with Urey's input designed glassware for the first experiments. The apparatus was simple, a continuously circulating system in which the simulated atmosphere, a mixture of CH_4, NH_3, H_2O, and H_2, is passed through sparking electrodes (mini-lightning bolts) that energize chemical reactions; the resulting brew is cooled, and newly made organics together with leftover gases collect in droplets that rain into the simulated ocean; and the ocean is gently heated to drive dissolved gases

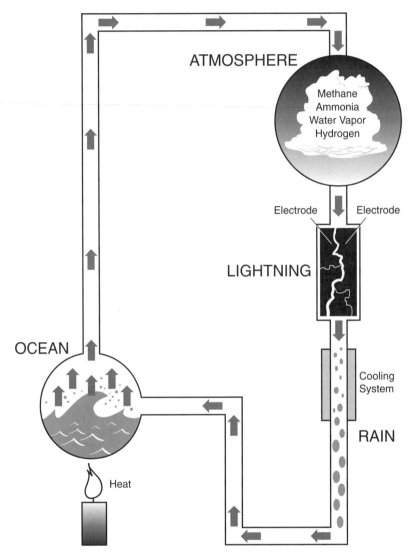

ATMOSPHERE

Methane
Ammonia
Water Vapor
Hydrogen

Electrode   Electrode

LIGHTNING

OCEAN

Cooling
System

RAIN

Heat

Figure 4.7 Apparatus used in Miller-type early Earth experiments.

back into the atmosphere to repeat the cycle. (Properly constructed, the apparatus poses no danger, but it must be absolutely airtight because air together with hydrogen or methane forms an explosive mix.)

After a few test runs, Miller turned on the spark and let it run for

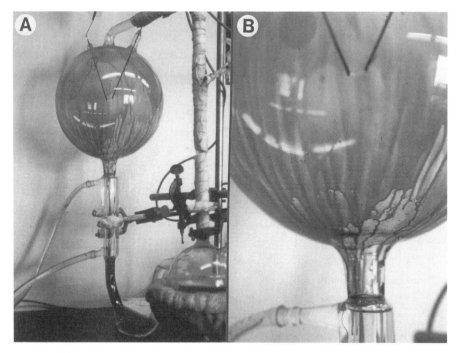

Figure 4.8 Amino-acid-containing brown organic sludge formed during the first 24 hours of sparking in a Miller-type experiment (**A**), shown close up in (**B**).

two days. The "ocean" became pale yellow. Boiling it down, he found evidence of glycine, the simplest amino acid of living cells. Urey was pleased, Miller ecstatic. They decided to repeat the experiment, this time letting the spark run for a week. In the final yellow-brown solution Miller detected seven amino acids including three (glycine, alanine, and aspartic acid) present in the proteins of living systems. Urey, too, was now ecstatic. In only three and a half months Miller had shown Urey's concept—and, evidently, Oparin's as well—to be correct. Moreover, Miller now had a first-rate thesis topic!

Publication of these findings a few months later (in May, 1953) caused a minor uproar. There even was a Gallup poll asking whether it was possible "to create life in a test tube" (9% said yes; 78%, no; 13% undecided).

These first experiments were mind-boggling and, to many, confusing. The claim was not that *life* had been made, but that *organic compounds* had. Yet in common parlance "living" and "organic" are almost inter-changeable and often equated with "natural" (as in organic farming and organic food in contrast to synthetic, "unnatural" products like chemi-cal fertilizers or pesticides). The distinctions are easily muddled, but the chemistry of life (biochemistry) is actually only one part of the much broader realm of organic chemistry—though both center on the chem-istry of carbon, especially in naturally occurring, CHON-containing compounds. The difference between them lies not in the chemicals made but in the processes by which the chemistry is done. Living cells are all composed of organic (carbon) compounds, and since these are made by biology they are biochemicals. But organic compounds, in-cluding many of the same ones produced by cells, are made also in the absence of life (in interstellar space, for instance, as we will see later in this chapter), so though these are organic and natural, they are not biological. The compounds made in Miller-type experiments are in this latter category—all are organic and many are identical to the bio-molecules of life, but none is of biologic origin.

Miller's pioneering success in making lifelike products from a nonlife process has been extended over the decades in laborato-ries throughout the world. Modified schemes have employed energy sources other than simulated lightning, using visible and UV light, cosmic radiation, heat (like that from volcanic rocks), shock waves (as though from meteoritic impacts), even the energy of ocean wave-fronts (released when tiny air bubbles are compressed by the force of breaking white caps). Because we now think the early atmosphere contained much less hydrogen than mimicked in Miller's early experi-ments, later ones have used gas mixtures containing only traces of molecular hydrogen and replaced methane with carbon monoxide (CO) or carbon dioxide (CO_2), ammonia with nitrogen gas (N_2). Still others have focused on the simple first-formed compounds—hydro-gen cyanide (HCN) and formaldehyde (H_2CO)—and explored atom-by-atom how these "reactive intermediates" interact to form amino acids and other monomers.

Alternative Ways to Turn the Trick

The search for a plausible scenario to explain life's origin is a work in progress, and though Miller-type syntheses have received most attention, now for more than 45 years, other models have sprung up. Since the amounts of organic matter made in early-Earth experiments are small when the simulated atmosphere includes only traces of hydrogen—about 10,000 times less than made in Miller's initial experiments—one model calls for the organic ingredients of the primordial soup to be delivered to Earth on impacting comets and meteoritic and other interplanetary debris (including interplanetary dust particles, "IDPs," that even today are estimated to pelt the Earth with a rain of hundreds of thousands of kilograms of IDPs each year). Another idea, thought promising just a decade ago, has it that life began as a "living information-containing clay mineral" that over time evolved to DNA-based organic life. Though this idea has fallen from favor for lack of experimental backing, the possible role of sheet-like clay minerals in life's origin is still actively pursued since the charged surfaces of clays can attract oppositely charged organic molecules and hasten their chemical interactions.

But the seemingly most promising alternate to the Oparin-Miller model is one proposed a few years ago by Günter Wächtershäuser, a chemist and patent attorney in Munich, Germany, according to which CO or CO_2 released from molten magma at deep-sea volcanic vents would have become stablilized on the surface of grains of iron-sulfur minerals where they could have reacted with molecular hydrogen in the vent waters to form organic monomers. This is a particularly interesting idea since it (1) fits with geologic evidence of abundant volcanism on the early Earth; (2) calls for organic compound synthesis in a setting rich in dissolved gases (potential starting materials), energy to power reactions (heat, or chemical energy from the reactions themselves), and varied temperatures and chemistries (giving diverse but fairly close-packed sites where different kinds of reactions might happen); (3) seems chemically plausible, since it relies on the kinds of reactions thought likely to happen under the conditions envisioned; (4) links the formation of monomers to a subsequent sequence of chemical events leading to life that though convoluted do not seem at odds with the basic biochemistry of living microbes; and (5) fits with evidence from rRNA evolutionary trees that some think shows early organisms lived in a high temperature setting.

However interesting, the idea is not without drawbacks. Though heat from deep-sea vents might power the chemical reactions it calls for, the same heat could destroy the monomers nearly as fast as made—for instance, at normal vent temperatures (about 300° C), even the sturdiest amino acids fall apart within hours and the fragile ones in minutes. And instead of the first forms of life being heterotrophs, the scenario envisions them to be autotrophs (technically "chemoautotrophs" since their metabolism is driven by chemistry, a term that separates them from plants and plantlike "photoautotrophs" that use light energy to power photosynthesis). So, they would have to have possessed biochemical machinery both to make food and break it down, a larger and much more complicated array of chemistry than needed by primitive heterotrophs.

The main drawback, however, is that the idea has yet to receive compelling experimental backing. Though at least seven laboratories are working actively on the problem (three in the U.S., two in Germany, one each in England and Russia), their findings have been less than spectacular. Evidently, the chemistry is not as straightforward as it looks in theory, and no one has been able to make the breakthrough that would give the idea firm footing, the crucial experiments showing that pyrite, FeS_2, can actually act as a matchmaker bringing hydrogen together with CO or CO_2 to build simple, biologically useful organic monomers such as amino acids (much less the larger compounds needed to mediate the reactions of chemoautotrophic metabolism). One glimmer of promise comes from studies that show carbon monoxide stabilized in boiling water on grains of iron and nickel sulfides combines with a four-atom organic fragment called a methyl group ($-CH_3$) to form acetic acid (CH_3COOH), a chemical that, once made, could react to form more complicated compounds. Though this is exceedingly simple chemistry—and though the compound produced, acetic acid, is not nearly as central to life as, for example, are the amino acids of Miller-type syntheses—the overall scenario continues to hold promise.

(As an aside, it's worth noting that this pyrite-mediated chemoautotrophic model seems to have garnered considerably more support among physical scientists, especially geologists and oceanographers, than among biologists, biochemists, and other life scientists. The interest of physical scientists is understandable since the scenario gives a prominent role to chemical reactions at deep-sea vents. But the relative lack of enthusiasm of life scientists is harder to fathom, though it may mean they think the problem's been solved, presumably by the Oparin-Miller heterotrophic hypothesis. Whatever the explanation, the fate of the new idea will hinge not on its popularity in opinion polls but on work in the laboratory. In the origin-of-life field, attractive theories are not nearly so scarce as the experimental backing needed to show they have counterparts in the real world.)

So, Miller-type nonbiologic syntheses may not be the only viable explanation for the buildup of small organic compounds on the lifeless Earth. Nevertheless, and though some alternate ways to turn the trick seem promising, it is the only model backed by a large body of solid experimental support. Indeed, scores of organic compounds have been identified in Miller-type experiments, including almost all common amino acids, the purine and pyrimidine bases of DNA and RNA, many types of hydrocarbons and fatty acids, and more than forty different sugars. Some protocols are more efficient than others, some more plausibly prebiotic, but all give biologiclike products if free oxygen is shut out of the system. The conclusion is inescapable—monomers of CHON *are* easy to make under simulated early-Earth conditions. But because the amounts produced would have been limited by the scarceness of hydrogen gas in the early atmosphere, organics formed this way were probably not plentiful and may have been augmented by other sources. Evidently, Darwin's warm pond and Oparin's primordial soup were close to the mark, though the soup may have been more a dilute consommé than a thick organic broth.

Organic Monomers beyond the Earth

It is a mistake to think Miller-type syntheses are aberrant, requiring some special type of chemistry carried out under extraordinary conditions. Only two prime ingredients are needed—CHON and the energy of starlight—and both are present throughout the Universe.

The universality of nonbiologic organic compound synthesis is shown by meteorites known as carbonaceous chondrites. Almost all meteorites are billions of years old, debris left over from when the solar system formed, and most are metallic, shiny mixtures of an iron-nickel alloy like that in the interiors of Earthlike planets. But others are rocky mineralic masses, evidently dislodged by violent collisions from the surfaces of the Moon and Mars and attracted by Earth's gravitational pull. And some are more crumbly objects such as the carbonaceous chondrites, so named because they are speckled with small glassy balls (chondrules) and charged with easily visible (as much as 5%) black coaly (carbonaceous) organic carbon. Unfortunately, rather few carbonaceous chondrites have been found because they break up on impact and rapidly weather away.

The origin and significance of the carbonaceous matter in these unusual meteorites has been of interest since the 1800s when its resemblance to coal, the compressed fossilized remnants of ancient plant life, seemed to hint at an extraterrestrial plant origin. But when various museum-stored specimens were examined, the biologiclike compounds (including amino acids) found were dismissed as contaminants. This picture changed on September 28, 1969, when a meteorite seen falling to Earth near Murchison, Australia, in the state of Victoria, was found to be a carbonaceous chondrite. Numerous pieces of the Murchison were picked up immediately and by high-powered techniques (gas chromatography–mass spectrometry) shown to contain seven amino acids, including five found in Miller's earlier studies. Further work identified eighteen amino acids in the meteorite and showed that these same amino acids, in almost identical relative amounts, are made in Miller-type experiments. This striking correspondence suggests that the early-Earth conditions simulated in Miller's experiments were present on the meteorite's asteroidal parent (a small planetlike body between Mars and Jupiter) and gave rise to the same suite of biologiclike monomers.

As the international space program has charted the heavens, abundant nonbiologic organic compounds have been found even farther afield—in the hydrogen-rich clouds encircling Jupiter and Saturn; in the atmosphere of Titan, Saturn's largest moon; in vagrant comets, such as the spectacular 1997 visitor Hale-Bopp; and even beyond the solar system. And though the radio telescopes of the SETI project (the Search for Extraterrestrial Intelligence) have not yet detected signals from distant, advanced, presumably carbon-based life forms, we do know that organic compounds are widespread in the Cosmos. Diverse

Figure 4.9 Two fragments of the organic-rich Murchison meteorite. The smooth surfaces are millimeter-thick fusion crusts formed as the firey meteor plunged through the Earth's atmosphere. (Bar for scale represents 1 cm.)

organic molecules in the gigantic dust clouds swirling through interstellar space have shown their telltale signatures in the microwave region of the electromagnetic spectrum. Discovered at a rate of about four compounds each year, more than eighty-five different kinds have been identified, the largest consisting of thirteen atoms (the hydrogen cyanidelike compound $HC_{11}N$). Their local abundance is often only one or a few molecules in a volume the size of a room. But because interstellar dust clouds are unimaginably vast, much larger than the solar system, the total number of molecules is astronomical.

All constituents of Urey's primitive atmosphere (CH_4, NH_3, H_2O, H_2) have been identified in these enormous clouds, as have many of the organic compounds pivotal in Miller-type early-Earth experiments: hydrogen cyanide (HCN), key for synthesis of amino acids and nucleic acid purine bases (adenine and guanine); thioformaldehyde (CH_2S), for sulfur-containing amino acids (cysteine and methionine); cyanoacetylene (HC_3N), for nucleic acid pyrimidine bases (cytosine, uracil, and thymine); formaldehyde (H_2CO), for monosaccharides including the ribose sugar of RNA; acetaldehyde (CH_3CHO), for the deoxyribose sugar of DNA; and cyanogen (C_2N_2) and cyanamide (H_2NCN), for forming the monomer-to-monomer linkages of polypeptides, polysaccharides, and polynucleotides; even methanol (CH_3OH, antifreeze) and ethanol (C_2H_5OH, the active agent of beer and liquor). All have been detected in prodigious quantities, in some interstellar clouds they have a total mass greater than the Earth.

The evidence is clear: organic monomers are widespread in the Universe, were abundant in the young solar system and present on the lifeless Earth, formed here by nonbiologic organic compound synthesis and delivered to the planet by impacting comets and interplanetary debris. But the buildup of such monomers under early-Earth conditions is only a first step toward uncovering the origin of life. The next hurdle is conversion of these simple chemicals into the polymers of living systems.

How Did Monomers Become Linked into Polymers?

Almost all of the key compounds of life are polymers, familiar large molecules such as carbohydrates, proteins, and the nucleic acids DNA and RNA. Though polymers can be huge, made up of millions of atoms, they are surprisingly uncomplicated. Nearly all are composed of an orderly sequence of much smaller, simpler, monomeric subunits.

A good example is cellulose, the crunchy substance of fresh vegetables (and the principal constituent of wood, the fabric of a cotton shirt, even the paper fibers on which this book is printed), the most abundant biomolecule on Earth, accounting for more than half the carbon in the living world. Though an exceedingly long molecule, made up of tens of thousands of carbon, hydrogen, and oxygen atoms, this carbohydrate is remarkably simple, a monotonous sequence of a single monomeric unit only six carbon atoms in size. The repeating unit is glucose ($C_6H_{12}O_6$), sugar monomers that are linked like beads on a string to form an almost endless chain.

Protein polymers are similar, threadlike aggregates of as many as twenty types of amino acid monomers linked in series. And nucleic acids are much the same, long polymeric strands made up of a regularly alternating sequence of sugar and phosphate monomers, with a

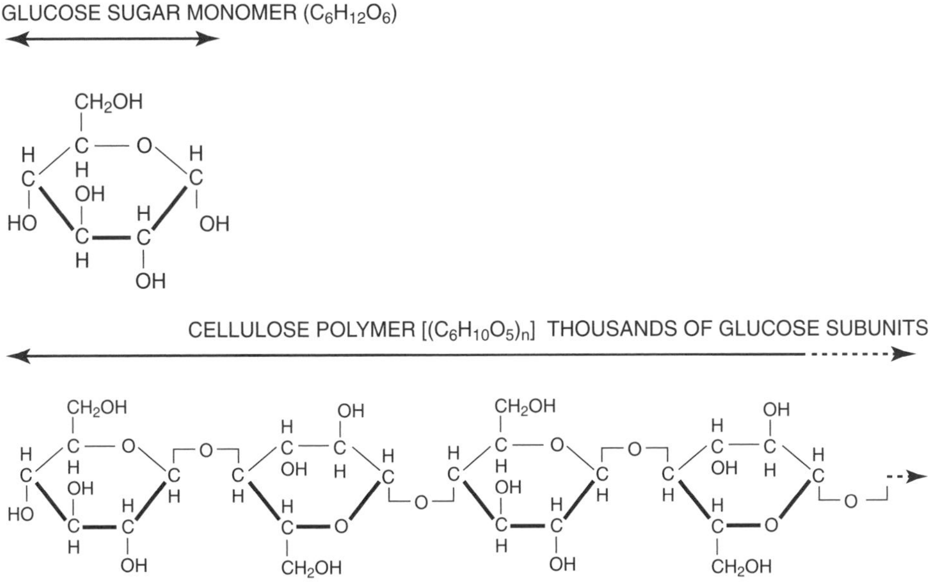

GLUCOSE SUGAR MONOMER ($C_6H_{12}O_6$)

CELLULOSE POLYMER [($C_6H_{10}O_5)_n$] THOUSANDS OF GLUCOSE SUBUNITS

Figure 4.10 Cellulose, a carbohydrate polymer.

purine or pyrimidine base attached to each sugar and bending outward from the sugar-phosphate backbone.

Cells build polymers by adding monomers one at a time to the end of a growing polymeric chain. To make the strong chemical bridge (covalent bond) that attaches each monomer to the lengthening chain, organisms use a process known as dehydration condensation. Consider, for instance, how amino acids polymerize to form a protein. All amino acids—even the simplest, glycine (H_2N-CH_2-$COOH$)—are short linear molecules, at one end having a cluster of three atoms, the amino group H_2N, and at the other a four-atom combination, COOH, the carboxylic acid group. Cells order amino acids head-to-tail in protein polymers by connecting the carbon of the COOH end of the growing polymer to the nitrogen of the H_2N end of each new amino acid added. But before the connecting bridge can be fit in place, space has to be freed on the carbon and nitrogen atoms. With the help of a bridge-building enzyme and powered by cellular energy, this is done by cutting away OH (hydroxyl) from the COOH of the polymer and excising one of the hydrogens from the H_2N of the amino acid, liber-

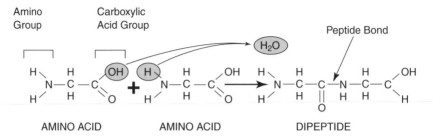

Figure 4.11 Dipeptide formation by dehydration condensation.

ated chemical fragments that then unite to form a molecule of water (OH + H → H₂O). This water-forming bridge-building process is repeated over and over as amino acids are added until the protein is completed.

Chemists term the linking C-N bridge a peptide bond, so the first pair of linked amino acids is called a dipeptide, and the terms polypeptide and protein have the same meaning. The technical name for the bridge-building reaction is dehydration condensation—*dehydration*, because water is removed as the peptide bond forms; *condensation*, because two molecules are combined (condensed) into one.

Carbohydrates and nucleic acids are made by the same process. Carbohydrates are polymers of sugar (therefore termed polysaccharides), and the sugar monomers that make them up have hydroxyl groups at each end of the molecule. By cutting away the OH at one end of a sugar monomer and combining it with an atom of hydrogen cut from the OH at the opposite end of an adjacent sugar, a molecule of water is formed as each monomer is linked to a growing carbohydrate chain. Nucleic acid biosynthesis involves three of these bridge-building dehydration condensation steps, two to form the nucleic acid building blocks (nucleotides) and a third to link these into a nucleic acid polymer.

Polymers of amino acids, sugars, and nucleotides, made in all organisms by dehydration condensation, are fundamental to life. This universality suggests the process is very ancient, probably dating from near life's origin. On the lifeless Earth, however, polymer-formation must have been simpler. Dehydration condensation requires removal

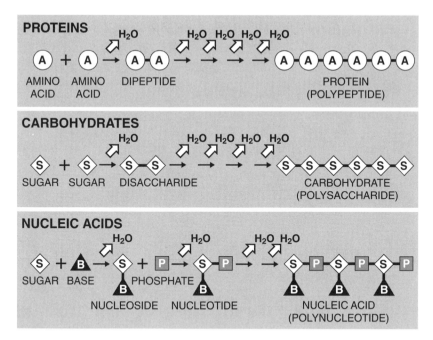

Figure 4.12 Formation of polymers from monomers by dehydration condensation.

of water from the combining monomers—a difficult feat in the watery setting where life began, like trying to dry one's hands with a sopping wet washcloth. Moreover, in cells today, this process is powered by cellular energy and speeded by complicated bridge-building enzymes, neither of which were present on the lifeless planet.

Three prebiotic ways to turn this trick have been devised. (1) Heated and dried as they would have been by contact with volcanic lavas, solutions of amino acids condense into proteinlike ("proteinoid") globules. (2) Chemically treated to enhance polymer-formation and spread on clays like those at the margins of shallow lagoons, amino acids polymerize into proteinlike compounds, as do nucleotides into nucleic acidlike products. (3) Mixed in water with small amounts of prebiotic "condensing agents" (such as cyanogen, cyanomide, and cyanoacetylene, formed in Miller-type experiments and present in interstellar space), sugars, amino acids, or nucleotides combine into biologiclike polysaccharides, polypeptides, or polynucleotides.

From Monomers to Polymers toward Life

The first steps to the emergence of life seem fairly clear. Powered by any of numerous energy sources, amino acids formed spontaneously (by what is known as a Strecker synthesis) from mixtures of formaldehyde, thioformaldehyde, ammonia, and hydrogen cyanide. Simple monosaccharides, including glucose and ribose, formed from the self-condensation of formaldehyde (by the Formose reaction); more complex sugars, such as deoxyribose, from acetaldehyde; purines from hydrogen cyanide; and pyrimidines from cyanoacetylene. From a dilute primordial soup of these monomeric ingredients, the prime polymers of life—proteins, carbohydrates, and nucleic acids—then self-assembled by prelife mechanisms of dehydration condensation such as those shown in the laboratory.

From these beginnings, cells evolved, but that part of the story, as we will see in the next chapter, is more hazy.

Metabolic Memories of the Earliest Cells

How Did Cells Begin?

One of the big puzzles of life's beginnings is how cells and their metabolism got started. One notion is that the first cells were perhaps like the smallest, simplest organisms alive today, microbes known as mycoplasmas. The cells of mycoplasmas are truly tiny, less than a billionth as massive as a protozoan, and they house only a fraction of the DNA and proteins normally present. But all mycoplasmas are parasites, scaled-down descendants of free-living larger microorganisms, and they grow and reproduce only inside other cells, usually in the innards of mammals, a lifestyle impossible for the earliest forms of life. An alternative model is an ordinary bacterium, but free-living microbes are much too complicated—made up of hundreds of different polymers (including about five hundred kinds of RNA), more than a thousand different enzymes, tens of millions of molecules. The first cells must have been vastly simpler.

To know what the earliest cells were like, we need to lift the evolutionary veil that separates life today from its beginnings. This task has only begun. Insight into life's early history comes one step at a time, and before the origins of monomers and polymers had been sorted out it seemed fruitless to tackle this daunting question. But though hardly anything is known for certain about how cellular life began, three stages in the sequence seem fairly clear:

1. Chemical systems most scientists would call alive probably existed before they became packaged in cells. Fossil evidence of this pre-

BACTERIAL CELL

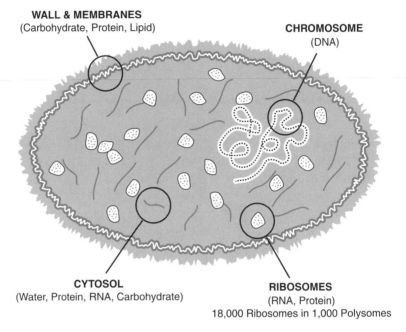

WALL & MEMBRANES
(Carbohydrate, Protein, Lipid)

CHROMOSOME
(DNA)

CYTOSOL
(Water, Protein, RNA, Carbohydrate)

RIBOSOMES
(RNA, Protein)
18,000 Ribosomes in 1,000 Polysomes

Figure 5.1 Principal components of a bacterial cell.

cellular stage may never be discovered since the chemicals are too fragile to be preserved, but for some period the early Earth is thought to have been populated by living molecules, nucleic acid "primordial naked genes" capable of self-reproduction.

2. Because of their chemical makeup, these and other organics of the primordial soup clumped together in tiny protocellular globules, likely ancestors of full-fledged walled cells. Evidence of this, too, seems lost to history though cells more ancient than the oldest now known may someday be unearthed.

3. The first cells were powered by a simple form of metabolism that laid the foundation for the evolution later of more complicated metabolic machinery.

We'll consider these three stages in turn.

The Chicken-and-Egg Problem

Life can multiply only if the chemical blueprint stored in the nucleic acids of chromosomes is copied and passed on to offspring. Most of this information directs the manufacture of enzymic proteins, a special kind of which (polymerases) is required to copy nucleic acids. So, which came first—the nucleic acids needed to make the polymerase enzymes, or the enzymes needed to make the acids?

For many years, attempts to solve this chicken-and-egg problem centered on DNA and its role in making enzymatic proteins. But because DNA itself is a product of evolution, the advanced, younger version of a more primitive ribonucleic acid ancestor, attention switched to RNA as the primal blueprint of life. In the early 1980s this refocus was answered by the discovery of ribozymes, a special kind of RNA that not only houses information but has enzymelike functions as well. Though all modern ribozymes are long and complicated, the short parts having enzymelike properties are simpler and may be more like the RNAs of early life.

Like enzymatic proteins, ribozymes can cut molecules apart or paste them together, and a number of them can do both. Some are self-splicers, able to snip away one part of their own length and glue the leftovers back together. Others can cut out a section of themselves and move it to another spot in the molecule. Still others can engineer assembly of fresh RNA strands. Though no ribozyme has been found to make a complete copy of itself, by repeated cutting and pasting many show an elementary capacity for self-reproduction. Experiments that pit various versions of these gene-enzyme hybrids against one another in test tubes even show how their evolution may have begun.

This series of recent discoveries makes it reasonable to imagine a precellular world in which primordial naked genes of RNA reproduced themselves without the help of protein enzymes. The ultimate goal of creating life in a test tube, from scratch fabricating "living molecules" that self-assemble and self-reproduce, is becoming less and less like science fiction.

Cells Are Like Bubbles of Soap

The need for cells in living systems is obvious. If the living juices of an organism are not walled off, they mingle with the surroundings and lose their order. This separation, perhaps not necessary for primal naked genes, was mandatory as life became complex.

The key to the origin of cells is the adage of chemistry teachers, "like dissolves like." Paint thinner and oily paint mix easily because they are chemically alike. But water and oil don't mix because their chemistries differ. Molecules of water are V-shaped, the two hydrogens of H_2O situated at the tips of the V linked to an atom of oxygen at the base. Because of their shape they behave like tiny bar magnets—at one end of the molecule each hydrogen carries a small positive charge, while at the other the oxygen carries a negative. Oil molecules, however, are uncharged, more like bits of plastic or wood. So oil balls up in water because its molecules dissolve more easily in each other ("like dissolving in like") than in the water around them.

Soaps are special compounds that bridge the chemical gap between water and oil. Charged atoms at one end of a soap molecule—the hydrophilic (water-loving) end—dissolve in water. The rest of the molecule is hydrophobic (water-fearing), a long chain of hydrogen and carbon atoms chemically like the oil and grease in which it dissolves easily. Soaps work because while one end mixes with water the other dissolves in grime.

Cells originated by soaplike chemistry. The primordial soup was a dilute consommé in which hydrophobic organic compounds clumped together naturally because of their chemical makeup. Among these were chains of hydrogen and carbon, hydrocarbons like the tails of soap molecules, some of which had charged atoms at one end. Like soaps, these packed together to make up thin-skinned bubbles in which the charged hydrophilic atoms formed the outer surface and the hydrocarbon tails pointed inward, mixing with the organics clumped inside.

How cells developed from this beginning can be seen, but only dimly. Though flimsy and fragile, the thin skins of the bubbles shielded the organics bunched within, allowing them to interact and form new configurations. Over time, a second sheet of soaplike mol-

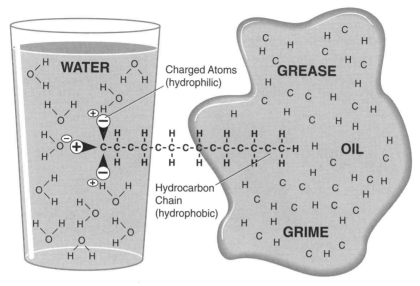

Figure 5.2 Soaps bridge the chemical gap between water and oil.

ecules combined back-to-back with the first to make up a two-layered bounding film, a structure like the membranes of modern cells in which the outer layer cordons off a compartment and the charged atoms on the inner surface enclose a watery mix of organic juices. This pliable bilayer then became strengthened by embedded proteins that helped maintain an exchange of nutrients and wastes with the surroundings, and later was reinforced by a robust carbohydrate-protein band that transformed it into a resilient enveloping capsule like the cell walls of present-day bacteria.

The physical structure, genetic blueprints, and earliest energy-making mechanisms of cells evolved in unison as life began, not one after another. But much of metabolism originated only later, millions of years after cells first appeared.

The Essentials of Life

Metabolism, the ways cells build foodstuffs and break them down, is surprisingly simple. Only a few ways were ever invented, almost all remodeled versions of what had come before. Metabolism rose like

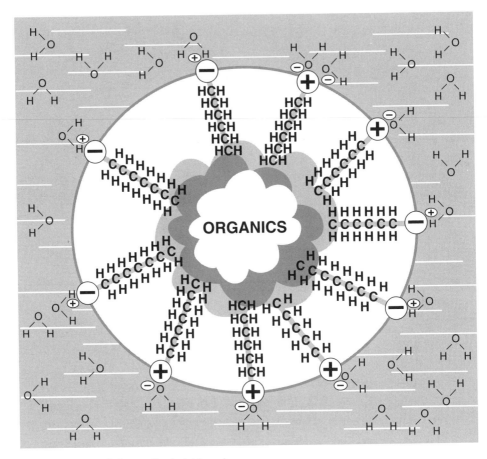

Figure 5.3 Cells are like bubbles of soap.

the ramparts of a medieval walled city put together in stages from building stones laid by earlier generations.

The history of metabolism is just one example of how evolution builds bit by bit on systems already in place. Why evolution operates this way is easy to understand. Cells are like incredibly intricate clocks, made up of a huge number of working parts that depend on one another in complicated ways. Big changes are likely to cause disaster. But minor tinkering can be accommodated, and tiny changes one after another add up. This Principle of Conservatism and Economy, an important take-home lesson about the evolutionary process, helps solve such interesting puzzles as why humans breathe oxygen and why life is divided into eaters and eatees.

Two million or so species of living organisms are known to science (with perhaps three to five times as many yet undiscovered). To stay alive, all need just two essentials—CHON and energy. Reproduction and evolution come with the package. Organisms can't evolve without reproducing, but that requires CHON, for the developing offspring, and energy as well.

Only two ways have been invented to meet these needs: autotrophy, the strategy of plants; and heterotrophy, the strategy of animals.

Autotrophs, or self-feeders (from the Greek, *autos*, self, and *trophos*, feeder), meet the need for CHON by taking in simple nutrients (usually carbon dioxide, water, nitrate, and phosphate) and building these into the foodstuffs that sustain them. Some autotrophs (called *chemo*autotrophs) gain the energy to build their foodstuffs from chemical reactions not powered by light, but most are photosynthetic (making them *photo*autotrophs), plants and plantlike microorganisms that grow by using the energy of sunlight to manufacture small organic compounds such as glucose. In both kinds of autotrophs, part of the energy harvested is stored in the chemical bridges made to link the atoms of C, H, O, and N, and can be released later and used by the organism. Autotrophs are self-builders *and* self-eaters—they take in CHON and make organic compounds, which they then break down to yield life-supporting energy.

Animals and animal-like organisms (protozoans, fungi, and most nonphotosynthetic microbes) have the other strategy, heterotrophy (from the Greek *heteros*, other, and *trophos*, feeders on others). Heterotrophs obtain CHON from the food they eat and energy from the bonds that link its atoms together. As Oparin saw decades ago, animals are metabolically simpler than plants. Heterotrophs use ready-made foodstuff and only need to break it down, but plants and other autotrophs must make their own food and then break it down as well.

The structure of today's living world is not complex. There are only two necessities, CHON and energy, and but two main strategies, photoautotrophy and heterotrophy. The living world has always had more or less this same simple structure, but it, like life itself, evolved

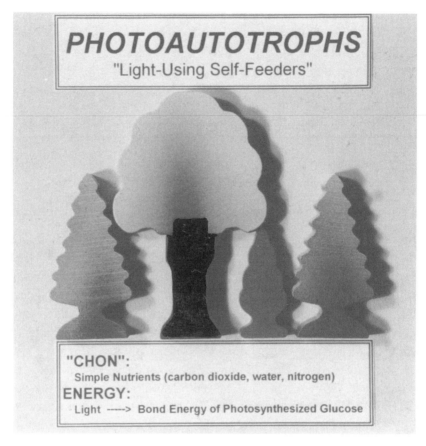

Figure 5.4 Photoautotrophs use the energy of light to make their own food.

from even simpler roots. Each of the two strategies comes in two versions, one primitive (early evolved) and one advanced (later evolved). The difference depends on whether molecular oxygen (O_2) plays a role. In the primitive form of photosynthesis, oxygen is not given off, and both it and the early-evolved form of heterotrophy are anaerobic, carried out in the absence of oxygen. But oxygen is central to the advanced versions of both—oxygen is generated in the later-evolved form of photosynthesis and consumed by advanced heterotrophs in the process of breathing.

Two necessities, two prime strategies, and only two versions of

Figure 5.5 Heterotrophs feed on others.

each strategy: one primitive, one advanced. Once invented by ancient microbes—long before plants or animals existed—this pattern was carried on to ecosystems everywhere. That's why the living world is divided into heterotrophic eaters and autotrophic eatees (and why we, like other heterotrophs, are so dependent on plant life). Evidently, evolution *is* remarkably conservative!

This is seemingly a good tale, but does it make sense? As we've seen in the rRNA Tree of Life, microbes are related only remotely to present-day plants and animals. How could their metabolic concoctions be passed along over billions of years and trillions of generations?

METABOLISM	"CHON" SOURCE	ENERGY SOURCE
ADVANCED - OXYGEN PRODUCED OR CONSUMED		
AEROBIC HETEROTROPHY Microbes, Protists, Fungi, Animals	**FOODSTUFFS PRODUCED BY AUTOTROPHS**	**AEROBIC RESPIRATION** O_2+GLUCOSE $\longrightarrow$ 36 ENERGY UNITS
AEROBIC PHOTOAUTOTROPHY Cyanobacteria, Protists, Plants	**OXYGENIC PHOTOSYNTHESIS** $CO_2 + H_2O \longrightarrow$ Glucose + O_2	**AEROBIC RESPIRATION** O_2+GLUCOSE $\longrightarrow$ 36 ENERGY UNITS
PRIMITIVE - OXYGEN NOT REQUIRED		
ANAEROBIC PHOTOAUTOTROPHY Photosynthetic Bacteria	**ANOXYGENIC PHOTOSYNTHESIS** $CO_2 + H_2S \longrightarrow$ Glucose	**GLYCOLYSIS** GLUCOSE $\longrightarrow$ 2 ENERGY UNITS
ANAEROBIC HETEROTROPHY Fermenting Bacteria	**FOODSTUFFS PRODUCED BY AUTOTROPHS (or primordial soup)**	**GLYCOLYSIS** GLUCOSE $\longrightarrow$ 2 ENERGY UNITS

Figure 5.6 Necessities and strategies of life.

Prepackaged Evolution

The answer is known, and the principal players are two groups of the Bacterial domain: cyanobacteria and purple bacteria.

The advanced form of photoautotrophy, oxygen-generating (oxygenic) photosynthesis, was invented by cyanobacteria, evidently as early as 3.5 billion years ago when the Earth was in its infancy. Much later, perhaps as recently as 2 billion years ago, one of their descendants was swallowed up by a single-celled eukaryote. Because the captured (but undigested) cyanobacterium carried with it the food-making machinery of photosynthesis, a mutually beneficial you-scratch-my-back-and-I'll-scratch-yours partnership was struck. The cyanobacterium served as a handy internal food factory, and the eukaryotic host provided shelter (a relationship we call "endosymbiosis").

Over time the alliance strengthened, most of the genes of the cya-

PREPACKAGED EVOLUTION

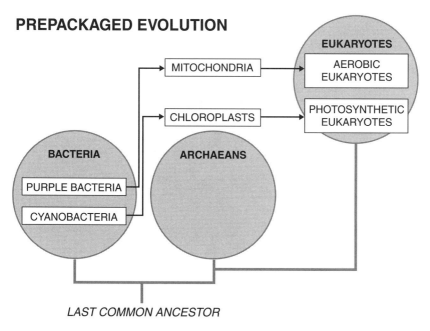

Figure 5.7 The endosymbiotic origin of chloroplasts and mitochondria.

nobacterium were transferred to the host, and the photosynthesizing endosymbionts evolved to become structures now known as chloroplasts, membrane-enclosed intracellular bodies (organelles) that house the photosynthetic apparatus in eukaryotic photoautotrophs such as plants. A similar sequence led to oxygen-breathing eukaryotes—endosymbiotic purple bacteria evolved to become the aerobic energy factories of eukaryotic cells, the rodletlike organelles known as mitochondria.

Evolutionary endosymbioses delivered food-making and air-breathing machinery prepackaged to eukaryotes, reusing systems already tested and perfected. The take-home lesson continues to hold—*evolution really is conservative and economical!*

Prepackaged evolution explains why the same type of photosynthesis is present in plants and cyanobacteria, but it does not account for its beginnings. Is it a recast version of some earlier invention or did it arise from scratch in cyanobacteria billions of years ago? And how did aerobic respiration get started? Breathing requires oxygen, so how could the process gain a foothold if there was no oxygen in the early atmosphere? In short, if evolution can only build on something that already exists, how do breakthrough advances ever happen?

Life's Earliest Way to Make a Living

Energy from Sugar Fermentation

Let's start at the beginning. Among the earliest forms of life were some that lived by glycolysis, a form of fermentation (anaerobic metabolism) in which a molecule of the six-carbon sugar glucose ($C_6H_{12}O_6$) is split in half to make two molecules of a three-carbon compound called pyruvate. This produces energy, given off when the chemical bonds of glucose are broken apart, some of which is stored for later use in a chemical known as ATP (adenosine triphosphate). Two units of energy (two energy-rich molecules of ATP) are made every time a molecule of glucose is broken down.

Glycolysis dates from near life's beginnings. It is fundamental to life, present in *all* organisms, a package of ten enzyme-speeded steps too large to have originated more than once. Moreover, it is chemically the simplest energy-making process in biology, takes place in the watery cytosol of cells (rather than needing membranes or organelles like later-evolved systems), yields much less energy than more advanced mechanisms, and is anaerobic like the early environment.

Glycolysis requires glucose fuel. But Miller-type early-Earth experiments show that many other sugars were present also in the primordial soup. Why was glucose pegged as the universal fuel of life? Probably because it is especially sturdy, the least susceptible of all six-carbon sugars to break down by changes in temperature, acidity, and the like. In the harsh early environment, glucose was the sugar most likely to be available to life.

By this (Oparin-Miller) scenario, the early Earth was populated by anaerobic heterotrophic microbes feeding on glucose of the primordial soup. Yet as these simple cells multiplied they would have begun to exhaust the fuel supply. Unless a new source of glucose turned up, they would have starved to the brink of extinction, and glycolysis might have been lost forever.

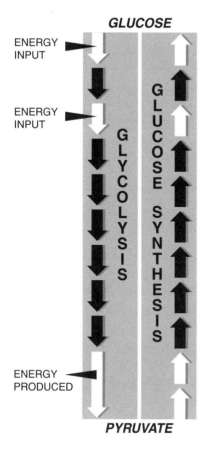

Figure 5.8 Glucose biosynthesis evolved by reversal of the ten-step process of glycolysis and replacement of three energy-mediating enzymes.

A New Source of Fuel

The problem was a shortage of glucose. To succeed, life had to find a better source. An early solution evolved in the form of microbes able to make glucose themselves using the glycolysis pathway remodeled to work in reverse. Manufacture of glucose (technically, "glucose biosynthesis") involves eleven enzyme-aided steps. Seven of these use the same enzymes as glycolysis but operate in the opposite direction. Without changing glycolysis in any way, genes for seven of its enzymes were duplicated and their enzymes used along with four new ones to construct the glucose-making system. Rather than inventing a brand-new set of genes and enzymes, *evolution was conservative and economical.*

How can the same set of enzymes help a sequence of chemical reactions go forward *and* backward, and why were only seven and not all ten of the glycolysis enzymes carried over to the new system for glucose synthesis? Imagine a toy train being moved along a track. If the track is level, the same force is needed to push the train forward as to pull it back. This is like most chemical reactions set in motion by enzymes—with equal ease the enzyme can push the reaction one way or pull it the other (that is, the reactions are reversible). But if the track goes over a hill, things differ. Extra energy must be added to shove the train up the hill, and energy (due to gravity) is released as it slides down the other side. In the same way, enzyme-aided reactions that either consume or release energy (pumped in or given off in the form of ATP) can be carried out easily only in *one* direction—they are in essence irreversible. In the breakdown of glucose by glycolysis, two steps require input of energy and a third makes energy for the cell. These three are the irreversible steps carried out by the four new enzymes invented for glucose biosynthesis.

This solution to the fuel shortage, however, would at best have been only a stopgap measure. Cells making glucose this way use up three times more energy than they gain when the glucose is broken down by fermentation. No organism can survive for long by consuming more energy than it makes. Like overspending on a credit card, this is a short-term solution bound to bring big trouble!

But glucose biosynthesis was a starting point that soon gave rise to a way of answering life's need for energy balance, the evolution of a new way to make glucose powered by light. First, however, another problem had to be solved.

Nitrogen Presents a Problem

Though glucose fuel is a ready source of carbon, hydrogen, and oxygen, life also needs nitrogen for proteins, nucleic acids, and ATP. What was the source of N in CHON?

Anaerobic heterotrophic microbes thrive if they are fed glucose (for CHO and energy) and ammonia (for N). Both were perhaps plentiful when life got started, but ammonia (NH_3) soon was in short supply. The chemical links that bind together the nitrogen and hydrogen in ammonia are broken easily by UV light. Because there was almost no molecular oxygen (O_2) in the early atmosphere, a UV-absorbing ozone (O_3) shield did not exist and ammonia was destroyed rapidly. Nitrate ($NO_3{}^-$), the other source of nitrogen used by many organisms, was also scarce. Today, large amounts of nitrate are made when oxygen and nitrogen combine during lightning storms, but this could not happen in the early oxygen-deficient atmosphere.

Because of the need for nitrogen, the scarcity of ammonia and nitrate posed a major problem to life. Only one other source was left: nitrogen gas in the atmosphere. But the atoms in N_2 are extraordinarily tightly linked, locked together by three strong bonds, so a special enzyme system was needed to break them apart. The agent that evolved to harvest nitrogen (to combine it with hydrogen and "fix" it in the form of ammonia) is called the nitrogenase or *Nif* complex, and its driving force is a protein called ferredoxin. Because N_2-fixation costs cells much energy, the *Nif* complex kicks in as a last resort, used only after supplies of ammonia and nitrate are exhausted. A system so costly would never have evolved had it not been crucial to life's survival.

The ferredoxin-driven *Nif* complex dates from early in Earth history when the environment was all but oxygen-free. Most early-evolved bacteria and archaeans can fix atmospheric nitrogen, whereas eukaryotes—all later-evolved—cannot, and like other especially ancient enzyme systems the *Nif* complex is brought to a standstill by trace amounts of molecular oxygen. N_2-fixation happens only if O_2 is shut out, even in oxygen-producing cyanobacteria, where special cells and chemical mechanisms have evolved to protect its workings.

How did ferredoxin originate? The fifty-five amino acids that make up the ferredoxin of a typical bacterium (*Clostridium*) are ordered in a way that reveals the history of the molecule. The protein started out as a snippet only four amino acids long. The gene for this quartet was copied repeatedly to make up a longer gene for a proto-ferredoxin composed of twenty-eight amino acids, seven of the quartets linked in a chain. Mutations then added an amino acid and switched several, and the mutated gene for the twenty-nine amino acid-long protein was duplicated to make a new gene for a primitive ferredoxin fifty-eight amino acids in length. After more mutations, three amino acids were cut away at one end of the molecule to give the ferredoxin of

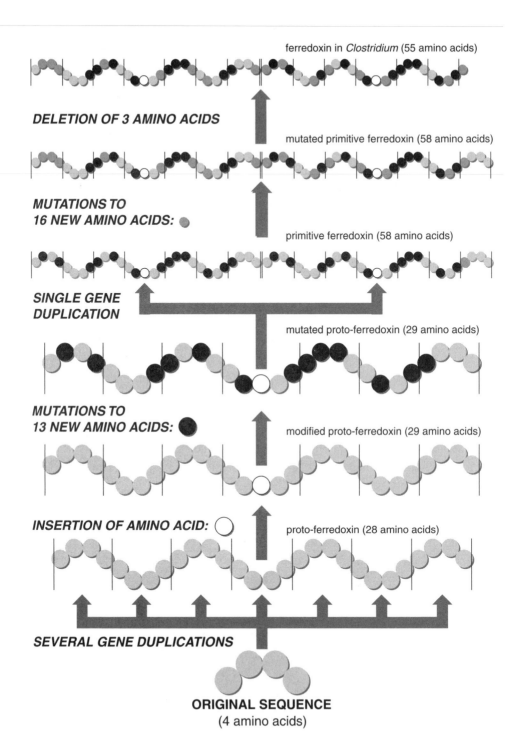

ferredoxin in *Clostridium* (55 amino acids)

DELETION OF 3 AMINO ACIDS

mutated primitive ferredoxin (58 amino acids)

**MUTATIONS TO
16 NEW AMINO ACIDS:**

primitive ferredoxin (58 amino acids)

**SINGLE GENE
DUPLICATION**

mutated proto-ferredoxin (29 amino acids)

**MUTATIONS TO
13 NEW AMINO ACIDS:**

modified proto-ferredoxin (29 amino acids)

INSERTION OF AMINO ACID:

proto-ferredoxin (28 amino acids)

SEVERAL GENE DUPLICATIONS

ORIGINAL SEQUENCE
(4 amino acids)

Figure 5.9 Evolutionary history of ferredoxin in the bacterium *Clostridium.*

modern *Clostridium*. Using as building blocks the initial amino acid quartet, ferredoxin evolved merely by copying and remodeling an already successful system. Once again, *evolution was conservative and economical.*

Gene xeroxing seems to have been especially common during life's early development, when CHON and energy were in short supply. Laboratory studies on experimentally starved bacteria show that almost all survivors are mutants that have extra copies of metabolic enzymes. Even in a normal setting, duplicated genes are present in about one in a thousand bacteria, so in a billion microbes, an ordinary-sized community for such minute forms of life, there are about a million that contain extra copies. As long as one copy of a gene functions in an organism, the extras can be mutated without harm, sometimes producing better genes that carry the blueprints for faster enzymes. As life got started, gene-copying made ample grist for the evolutionary mill.

Air and Light: A New Source of Glucose

By the scenario outlined so far, the earliest strategy of life was anaerobic heterotrophy—CHO was gained from glucose of the primordial consommé, N from atmospheric nitrogen (and ammonia and nitrate, where available), and energy from fermentation. But glucose fuel was scarce, even after cells found a way to make it themselves. This problem was solved by the appearance of autotrophs capable of photosynthesis, a process that makes generous amounts of glucose (unlike nonbiological mechanisms) in an energetically cost-effective way (unlike glucose biosynthesis).

There are two forms of photosynthesis—one primitive, one advanced—and they have much in common. They use similar pigments and chemistries to build the same product (glucose) from the same starting material (carbon dioxide) by practically the same pathway (glucose biosynthesis powered by light), and both are present in members of the Bacterial domain. The primitive photosynthesizers, varieties of photosynthetic bacteria, use bacteriochlorophyll to harvest light and do not produce oxygen as a photosynthetic by-product (that is, the process is anoxygenic). Advanced photosynthesizers, cyanobacteria,

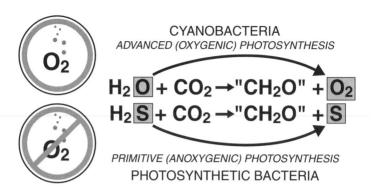

Figure 5.10 Whereas oxygen is a by-product of advanced, cyanobacterial photosynthesis, it is neither given off nor tolerated by the primitive photosynthetic bacteria from which they evolved.

use chlorophyll (of the same kind present in plants) to capture light energy, and by oxygenic photosynthesis they *do* give off oxygen. Both types of photosynthesizers can also use light energy to aid uptake of ready-made foodstuffs from the environment, a lifestyle known as photoheterotrophy and even more primitive than either type of photosynthesis. So, light energy was apparently first used to aid heterotrophy and only later recruited to power the glucose biosynthesis of autotrophic photosynthesizers—*another example of the conservatism and economy of evolution.*

Though closely related, the two forms of photosynthesis differ. Both combine hydrogen with carbon dioxide to build glucose, but the hydrogen comes from different sources. In the primitive process, hydrogen is supplied by hydrogen gas (H_2), small organic compounds, or hydrogen sulfide (H_2S). In advanced photosynthesis, hydrogen is always provided by water, and for this reason oxygen is given off, freed from H_2O as the hydrogen atoms are split away. Use of hydrogen gas or the hydrogen of organic compounds does not require much energy and separation of hydrogen from H_2S only somewhat more (78 kilocalories). But hydrogen and oxygen are knit together tightly in molecules of water, and to tear them apart requires much more energy (118 kilocalories).

Primitive, anoxygenic photosynthesis uses less energy, the advanced process more. And the primitive system is simpler by having only a single light-sensitive photosystem where light energy is captured and used whereas the advanced has two, linked by a string of enzymes that pass energy one to the other. Almost always, evolution is conservative and economical. Yet advanced oxygen-producing photosynthesis uses more energy and is more complex than the earlier evolved process. What offset these shortcomings?

Microbial Gas Warfare

Part of the answer lies in the availability of hydrogen. In forms usable by primitive photosynthesizers, hydrogen was abundant locally—in hot springs and fumaroles, for example, where H_2S bubbles to the surface—but elsewhere it must have been rare. Water, on the other hand, was available almost everywhere, and its use by advanced photoautotrophs allowed them to spread into pristine locales. Interactions with molecular oxygen mattered even more. Whereas slight amounts of free oxygen switch off the enzymes used by primitive photosynthetic bacteria to fix nitrogen or manufacture bacteriochlorophyll, advanced oxygen-producing cyanobacteria thrive in its presence. This difference had an enormous impact on the history of life.

Imagine what happened when the first oxygen producers entered the scene. The microbes of this new mutant strain, the first cyanobacteria, shared a shallow-water environment with their parental stock, anaerobic photosynthetic bacteria, where the two groups competed side by side to harvest sunlight. But the newcomers brought a telling advantage to the Darwinian struggle. Their new type of photosynthesis produced molecular oxygen, a gas toxic to their anaerobic neighbors. Microbial gas warfare! The bacterial parents were in dire straights, unable to fix nitrogen gas or build bacteriochlorophyll. Their survival in jeopardy, they could only retreat or die.

Thanks to tricks learned earlier in their history, the hard-pressed parental stock did retreat and did survive. When these primitive photosynthethic microbes originated and first evolved, the world was practically devoid of free oxygen and bathed in deadly UV light. To photosynthesize they had to see the sun, but if they tried to grow where there was too much light they would be fried alive. To cope, they lived on the murky seafloor, shielded by a cover of water, and many developed the ability to glide away from intense sunlight. When

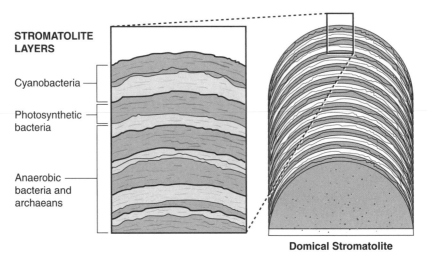

STROMATOLITE LAYERS

Cyanobacteria

Photosynthetic bacteria

Anaerobic bacteria and archaeans

Domical Stromatolite

Figure 5.11 Zoned by gradients of light and oxgen, stromatolitic microbial communities have cyanobacteria at their upper surface, photosynthetic bacteria in the layers beneath, and dark-dwelling anaerobic bacteria and archaeans below.

oxygen-spewing cyanobacteria invaded the scene, anaerobes unable to move no doubt died in droves, but photosynthetic bacteria weathered the storm by retreating to an oxygen-free setting within the seafloor muds. Primitive (but clever!) anoxygenic photosynthesizers survived by fleeing the war zone.

Today, cyanobacteria and photosynthetic bacteria live harmoniously in layered stromatolitic communities, able to coexist because they have different light-capturing pigments. As we will see in chapter 7, the oxygen producers populate the uppermost layer, anoxygenic photosynthesizers the layers below (and anaerobes that can live without light, the deepest stromatolitic zone). Though most of the sunlight is soaked up by cyanobacterial chlorophyll, this does not snuff out the underlying bacterial photosynthesizers because their bacteriochlorophyll is sensitive to wavelengths of light that seep through to where they live.

Why Do We Breathe Oxygen?

An old saying has it that "one person's trash is another person's treasure." To anaerobes, oxygen is not only other organisms' waste, it is a deadly poison. But to aerobes the reverse is true—oxygen is an elixir, the essence that powers the process on which their very life depends, aerobic respiration or "breathing." How does this work and how did *it* come to be?

Aerobic respiration has three parts. First, *glycolysis* breaks down glucose to make pyruvate and two ATPs (and water) for every molecule of glucose used. Second, the pyruvate is split apart by a cyclic system (the *citric acid cycle*) to form two more ATPs, electrons, and carbon dioxide. Third, molecular oxygen is pumped in and electrons from the citric acid cycle are conveyed along a string of enzyme-driven *electron carriers* to produce thirty-two more ATPs. Overall, thirty-six ATPs are made from each molecule of glucose broken down.

Breathing oxygen via aerobic respiration is a vast improvement over earlier-evolved glucose fermentation (glycolysis alone). The primitive process makes two ATPs from each glucose metabolized— the equivalent of only 2% of the energy stored in each molecule— whereas the oxygen-using system yields thirty-six, a whopping 38% of the energy available (and a yield better than the 25% efficiency of most automobile engines).

What are the evolutionary roots of this life-sustaining, notably cost-effective process? The first part, glycolysis, is already familiar. Inherited from primitive anaerobic heterotrophs, it long predates the appearance of oxygen-breathing forms of life. The second, the citric acid cycle, is also a hand-me-down, the borrowed but reversed version of the cyclic "dark reactions" of bacterial photosynthesis. And the third, the oxygen-consuming part of the system, is a revamped form of the chemistry that links the two light-sensitive photosystems in oxygen-producing photosynthesis. By remodeling and reusing inventions perfected earlier, evolution once again was *conservative and economical.*

The evolutionary roots of aerobic respiration can be seen even in children at play. When youngsters run fast and hard, they sometimes get "side aches" and have to rest to catch their breath. Why does that stop the pain? Side aches happen when muscles use oxygen so fast that not enough is left for aerobic respiration. As the oxygen debt builds, glycolysis makes more pyruvate than can be broken down, and the excess is converted to ache-causing lactic acid. Catching one's

AEROBIC RESPIRATION

OVERALL REACTION:

$$C_6H_{12}O_6 + 6\,O_2 \longrightarrow 6\,CO_2 + 6\,H_2O + 36\ \text{ENERGY UNITS}$$

glucose   molecular   carbon   water   cellular energy (ATP)
sugar   oxygen   dioxide

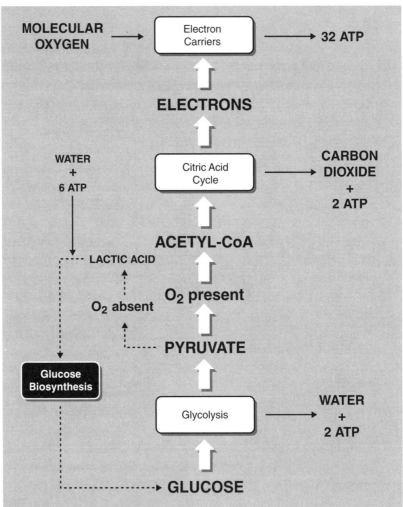

Figure 5.12 The metabolic pathway of aerobic respiration, "breathing."

breath helps restore the balance between oxygen inflow and pyruvate production. Given enough time, the pooled lactic acid is converted to glucose (by the glucose biosynthesis pathway) and cycled back into respiration.

Vigorous exercise causes humans and all other animals to revert to the most primitive forms of metabolism—glycolysis and glucose bio-synthesis—processes invented by anaerobic microbes in the distant geologic past.

The Four-Stage Development of Modern Metabolism

The metabolism that drives the modern ecosystem evolved in four stages, all but the first recast versions of what had come before:

1. Anaerobic Heterotrophy (Fermentation)

Among the earliest ways to live was by glycolysis, a simple chemical pathway used by anaerobic heterotrophs to generate energy by breaking down glucose taken in from the environment. As glucose supplies became used up, cells invented a reversed version of glycolysis, the glucose biosynthesis pathway. This made more glucose fuel, but worked only as a stopgap measure since it cost more energy to carry out than it produced. Glucose remained in short supply. Nitrogen in forms usable by early life was also scarce, a problem overcome by invention of ferredoxin-driven N_2-fixation. But this, too, was energetically costly.

2. Anaerobic Photoautotrophy (Anoxygenic Photosynthesis)

Light energy harvested by pigments newly evolved in anaerobic heterotrophs improved the means for uptake of organic compounds from the surroundings. Modification of this bacteriochlorophyll-based machinery allowed CO_2 to be used in place of organics as the source of cellular carbon, and when linked to the earlier-evolved glucose biosynthesis pathway it marked the origin of anoxygenic photoautotrophy, an early form of photosynthesis in which light energy captured in a single photosystem is used to build glucose, the universal cellular fuel. This was cost-effective and provided copious glucose,

EVOLUTION OF METABOLIC PATHWAYS

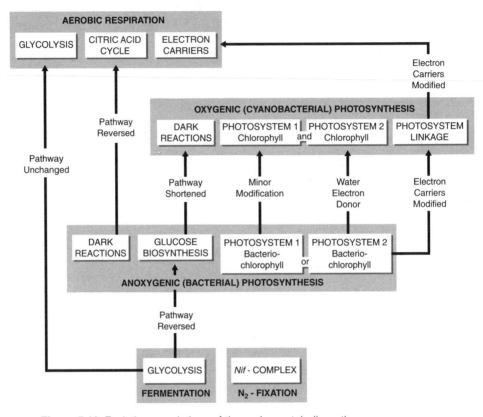

Figure 5.13 Evolutionary relations of the major metabolic pathways.

freeing life from dependence on nonbiological foodstuffs, but it relied on hydrogen sources that were plentiful only locally.

3. Aerobic Photoautotrophy (Oxygenic Photosynthesis)

A more complicated form of photoautotrophy evolved next, based on two chlorophyll-containing, light-sensitive photosystems linked together and using an especially abundant source of hydrogen: H_2O. This new water-splitting cyanobacterial form of photosynthesis gave off oxygen, a gas toxic to competing anaerobes. Unable to cope, some anaerobes went extinct while others retreated to more tolerable settings, leaving vast stretches of the world's surface open to occupation by cyanobacteria, the new rulers of the realm.

4. Aerobic Heterotrophy (Aerobic Respiration)

The increasing amount of molecular oxygen pumped into the environment by cyanobacterial photosynthesis presented a cardinal opportunity to life. Energy is released rapidly when oxygen combines with organic substances (as in wildfires, for example), a rich energy store tapped by the final metabolic invention, the three-part process of aerobic respiration. The first part, glycolysis, was borrowed unchanged from fermenting anaerobic bacteria. The second, the electron-generating citric acid cycle, is a recast version of chemistry invented by photosynthetic bacteria. And the third, is an oxygen-consuming pathway modified after the electron carrier system of oxygen-producing photosynthesizers. By tying these parts together, evolution devised the process of breathing, a powerful new way of living.

The Ancient "Modern" Ecosystem

The key metabolic processes of life today—heterotrophy and photo-autotrophy, anaerobic and aerobic—were all invented by microbes billions of years ago. The same systems are used, the same rules apply, whether CHON and energy are cycled between animals and plants as they are today or only among microorganisms as in the distant past. The present-day ecosystem is not modern at all but merely the scaled-up version of one first put in place by a menagerie of ancient microbes!

6

So Far, So Fast,
So Early?

How Old Is the Modern Ecosystem?

The oxygen-based lifestyles that so dominate in the living world today are founded on ancient anaerobic footing, oxygen-producing photoautotrophy on anoxygenic photosynthesis, oxygen-consuming respiration on anaerobic fermentation. We can be sure that once oxygen producers and oxygen consumers were added to the primordial anaerobic world, the basic workings of the modern ecosystem were set in place. To ferret out the antiquity and trace the history of this ecosystem we look to three lines of evidence: (1) the fossil record of cyanobacteria—the earliest-evolved "complete aerobes" capable both of producing and consuming oxygen; (2) minerals of ancient rocks that can tell us how much oxygen was in the atmosphere; and (3) chemical signals preserved in sedimented organic matter that show whether photosynthesis was going on. All lines point to the conclusion that oxygen-based metabolism—and with it, the modern ecosystem—arose only a few hundred million years after life got started.

Evidence from the Oldest Fossils

As we saw in chapter 3, the oldest fossils known—tiny cellular threads entombed in the Apex chert of northwesternmost Australia—are nearly three-quarters the age of the Earth. The fossil-bearing bed was laid down along the edge of a narrow seaway flanked by soaring volcanoes that episodically blanketed the seafloor with massive lava flows.

Sandwiched between two of these, each precisely dated, the fossiliferous horizon is 3,465 ± 5 Ma old.

In view of their fragile makeup, minute size, and immense age, the fragmented carbonized Apex fossils are surprisingly well preserved. Like the growth rings of fossil tree trunks, they are petrified, composed of coaly remnants of their original cell walls, the living juices leached away long ago and the cells filled in with mineral. Practically all that is left to divulge their identity is the size, shape, and boxcarlike arrangement of their simple filament-forming cells. But these clues from some 1,900 cells in hundreds of examples of the eleven kinds of fossils found reveal also how the cells divided and multiplied and give good reasons for judging some to be cyanobacteria.

Several types closely resemble living cyanobacteria of the taxonomic family Oscillatoriaceae, today an especially common group shown by rRNA evolutionary trees to be one of the most primitive kinds. Their relation to this particular family of cyanobacteria also meshes with the younger, Proterozoic, Precambrian fossil record, where oscillatoriaceans preserved in the same way in similar settings and microbial communities are widespread and abundant (as we will see in chapter 8). Moreover, as we will learn later in this chapter, it fits well with the chemistries of ancient minerals and organic carbon, which show that cyanobacterial photosynthesis may actually date from even earlier than 3.5 billion years ago. Cemented to rocks and boulders on the seaway floor by a thick layer of sticky mucilage and protected by overlying waters from lethal UV light, the Apex community was evidently composed of cyanobacteria as well as other kinds of prokaryotes, and included autotrophs and heterotrophs and both anaerobes and aerobes (the last, probably "facultative" like some oscillatoriaceans and many other microbes that consume oxygen when available but otherwise live by anaerobic means).

Even under the best circumstances, tiny delicate microorganisms like those of the Apex chert stand an exceedingly slim chance of making it into the fossil record and then surviving to the present. Yet survive they did, and though they give only a scanty glimpse of the ancient living world, the preservation even in a single fossil horizon of such a varied microbial community is a clear indicator that a profusion of microorganisms was present at this very early stage in Earth history. By the time of the Apex chert, diverse microbes, many like those today, were already thriving not only in what is now Western Australia but in practically all habitable settings across the globe.

When Did Life Begin?

The origin of life cannot be dated precisely. The Apex fossils set a minimum age, yet are too varied and advanced to be close to life's beginnings. Living systems arose earlier than 3,500 Ma ago, during the first billion years of Earth's existence, but just when is an open question.

Because of ordinary geologic processes—uplift, erosion, plate tectonics, and the recycling of older rocks to younger—only one sequence of strata more ancient than the Apex-containing package has survived. This set of deposits, the greatly pressure-cooked Isua Supracrustal Group of southwestern Greenland, dates from about 3.8 billion years ago. Organic carbon in the Isua rocks has been converted to crystalline graphite, and though some scientists have supposed it to be a residue of early (possibly photosynthetic) life, it could equally be charred dregs of primordial soup, the remains of nonbiologic organic matter formed on the early Earth or brought in with meteorites or comets. The Isua graphite hints that life may have existed, but its chemistry is too altered to serve as a smoking gun telling us how and when life began or for certain even that life was present. Rocks are unknown from the earlier seven hundred million years of the planet's existence. There is no older earthly evidence to reckon with.

What happened during the formative phase of Earth history, the hundreds of millions of years missing from the geologic record? The Moon holds the answer. The Moon is too small to have plate tectonics or own an atmosphere. There is no mountain building, it never rains, and rocks don't weather like they do on Earth. Though churned by meteorites over the ages, the otherwise pristine lunar surface retains a record of its embryonic development. The scarred and cratered moonscape reveals what a horrendous time this was.

During the segment missing from our planet's geology, the Moon, the Earth, and other bodies of the solar system were in the final stages of formation, sweeping up huge chunks of rocky debris encountered in their orbits. They were bombarded, blasted, by infalling meteors. And because the Earth is so much larger than the Moon, the barrage here was much greater. The heat of the bombarding fragments melted out the Earth's iron core, which settled to the center of the forming planet. Lighter material floated on its seething molten surface and by 4,200 Ma ago had begun to congeal into a rocky rind. So intense was the bombardment and so enormous the infalling meteors that the Earth's oceans were boiled away repeatedly. Even a medium-sized impactor 5 kilometers across would have turned the oceans to steam, shrouding the planet in enormous, lightning-charged, pitch-black cloud banks that would have lasted thousands of years before they rained out to reform the oceans. Had any life gained a foothold in this hellish setting, it would have been wiped out as the entire planet was sterilized over and over again.

Catastrophic collisions have happened throughout the history of the planet and are sure to happen again, though hopefully not soon and never with the devastation that battered the early Earth. Sixty-five million years ago a rocky mass 10 kilometers across smashed into the Central American Yucatán Peninsula, bringing tidal waves, fire, and soot that decimated the world's ecosystem and helped to wipe out the dinosaurs. In 1908, a much smaller 70-meter body blew up over Tunguska, Siberia. The 20-megaton blast ignited the clothes of a man almost 100 kilometers away, flattened thousands of square kilometers of densely packed forest, and annihilated reindeer to a distance of 50 kilometers. Every 100,000 years or so, Earth is hit by a kilometer-sized meteorite that throws enough debris into the atmosphere to block out the sun, plunging the world into a years-long night, nature's equivalent of a "Nuclear Winter."

But terrifying events even such as these pale to insignificance in comparison with the deadly storm that rained onto the primal Earth for hundreds of millions of years. The sterilizing devastation kept on until about 3,900 Ma ago, when the last of the huge orbiting chunks was swept away. Living systems may have originated and been killed off many times during earlier planetary history—there is no way to tell. But life as we know it could come into being only after 3,900 Ma ago, and by a scant 400 Ma later it was flourishing and widespread. How did life advance so far so fast?

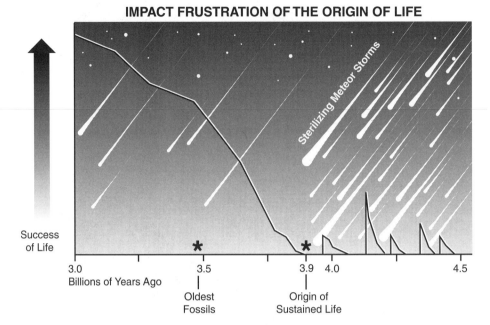

IMPACT FRUSTRATION OF THE ORIGIN OF LIFE

Figure 6.1 The common ancestor of life today could have originated only after the last great planet-sterilizing meteoritic impact.

How Did Evolution Proceed So Far, So Fast, So Early?

There are no fossils to give hard evidence of why evolution was evidently so speedy between 3,900 and 3,500 Ma ago. Though the goal of science is to know, not just to speculate, we can only make an educated guess. But we have one card up our sleeve. The fossil record of later geologic time should reveal how much evolution can happen in a 400-Ma-long period.

What was life's lot during the most recent 400 million years? First, primitive land plants appeared, tiny twigs less than a centimeter high. Rising from these unimpressive beginnings, plant life over the following 400 million years evolved to produce lush lowland marsh vegetation, giant spore-producing trees of the Coal Swamp Floras, highland seed plants, luxuriant conifer forests, and ultimately all of the trees, shrubs, and flowering plants of the modern world. While this was going on, *all* animal life on land evolved as well—amphibians (salamanders, frogs, and the like), reptiles (including dinosaurs), and warm-blooded birds and mammals, including—almost as an afterthought—primates such as ourselves. An enormous amount of evolution can be squeezed into 400 million years!

But the impressive feats of recent geologic time may not be a fair test of the speed of early evolutionary advance. By 400 Ma ago, life had long been in place, yet if organisms existed 3,900 Ma ago they were primitive and simple. How could embryonic life evolve so far and fast?

The mechanisms of early evolution may hold the key. As we saw in chapter 5, gene copying was rampant during life's early stages. As long as one copy of a gene functions, the extras can be mutated, sometimes to genes that work even better. And because new genes don't have to be made from scratch, this is a surefire mechanism for rapid evolution. Metabolism evolved by a similar pattern. The critical inventions in early biologic history were ways to meet the needs for CHON and energy. But rather than concoct totally new ways to do these jobs, evolution took shortcuts by remodeling and reusing those already in place. Gene xeroxing and evolutionary remodeling—both examples of the conservatism and economy of the evolutionary process—played major roles in speeding life's early advance.

Paleobiology: Fossils, Geology, and Geochemistry

From its beginnings in the early 1800s, Phanerozoic paleontology has focused almost entirely on fossils—their body forms, life histories, habitats, evolutionary relations, and spread in time and space. Extension of life's history into the Precambrian raised new challenges that have demanded a different perspective and a broadened plan of attack.

Practically all Phanerozoic fossils belong to groups living today, so questions about their biochemistry and metabolism rarely arise. Even extinct forms are immune—no one wonders whether dinosaurs were oxygen-breathing or the earliest land plants capable of photosynthesis. Yet in the Precambrian, such questions are central. Much of early evolution hinged on the development of metabolism. But the timing and trends of early metabolic evolution are difficult to sort out by traditional fossil-focused paleontology, largely because microbes of similar size and shape can differ greatly in metabolic lifestyle.

In the same vein, though major changes in the global environment are mostly too gradual and long-term to affect the Phanerozoic history of life, the Precambrian encompasses such an enormous sweep of time that even accumulated minor changes can have a telling effect. For example, the 21% oxygen of today's atmosphere would have been generated by the steady addition of only a tiny amount (0.000000006% O_2 per million years) from the time of the Apex fossils to the present. Yet this same rate of increase from the beginning of the Phanerozoic would result in the rise of oxygen to only 3%, a level too low to sustain all but the most primitive life around us.

The fresh questions raised by the Precambrian have begun to be answered by "paleobiology"—a term evidently coined in 1904 by Yale paleontologist Charles Schuchert—a broad new science that combines traditional paleontology with the evidence and insights of geology and isotopic geochemistry.

Geologic Evidence of Oxygenic Photosynthesis

Of all microbes, only cyanobacteria carry out oxygen-producing photosynthesis, a process that of course cannot take place without its two starting ingredients, water and carbon dioxide, and that forms two products, organic matter and molecular oxygen. Evidence of all four should be unmistakable in the rock record once cyanobacteria appeared.

There is ample proof early of liquid water (see plate 5). The fossil-bearing Apex chert is part of a sequence of volcanic and sedimentary rock some 15 kilometers thick, nearly all deposited in shallow seas. Its geology tells us the scene was dominated by sinuous seaways with marginal lagoons and scattered volcanic islands fringed by water-laid gravels, sands, and muds. The other ingredient needed for photosynthesis, CO_2, was also plentiful. Carbon dioxide solubilized in seawater as bicarbonate (HCO_3^-) combined with dissolved calcium (Ca^{+2}) to form the calcium carbonate ($CaCO_3$) minerals that make up limestone beds of the Apex sequence.

The two end products of cyanobacterial photosynthesis, organic matter and O_2, were present as well. Particles of coaly carbonaceous matter, kerogen, are abundant in the Apex rocks (constituting up to about 1% of their weight) and make up the carbon-rich cell walls of the petrified fossil microbes. The presence of free oxygen is shown by iron-oxide-rich banded iron formations (BIFs). Abundant in terrains older but not younger than about 2,000 Ma, these and other geologic indicators reveal much about the early history of the atmosphere.

BIFs and the Rusting of the Earth

BIFs are the principal source of the world's iron ore. Their distinctive millimeter- to centimeter-thick banding is caused by alternation of iron-rich and iron-poor layers; because the iron is in the form of tiny rustlike grains of hematite (Fe_2O_3) and sometimes magnetite (Fe_3O_4) they have a telltale bright to dull red color (see plate 6).

The iron owes its origin to the circulation of seawater through hot cracks and fissures in the ocean floor, primarily at the deep submarine ridge systems that play an active role in plate tectonics. In a dissolved form (as ferrous, divalent iron, Fe^{+2}) it then spread upward into shallower reaches of the water column—often seasonally, giving rise to the distinctive fine banding. There it was oxidized to the ferric state (trivalent iron, Fe^{+3}), chiefly by combination with dissolved molecular oxygen, and rained out of solution as a rusty mist of minute iron oxide particles.

Except for gases given off by organisms, the atmosphere is a product of geologic processes, the accumulated vapors exhaled by volcanic venting of the Earth's interior over geologic time. Unlike the atmosphere's other major constituents (N_2, H_2O, CO_2), molecular oxygen is not discharged from rocks when they are heated. But an enormous amount of oxygen—more than 20,000 million trillion grams (2×10^{22} gm), roughly twenty times as much as in the atmosphere today—is buried in the iron oxides of BIFs. Of all sources, including nonbiologic ones (such as the breakdown of water vapor by intense heat or high-energy UV light), only photosynthesis is capable of generating such a massive amount.

BIFs are abundant over a long span of geologic time, from earlier

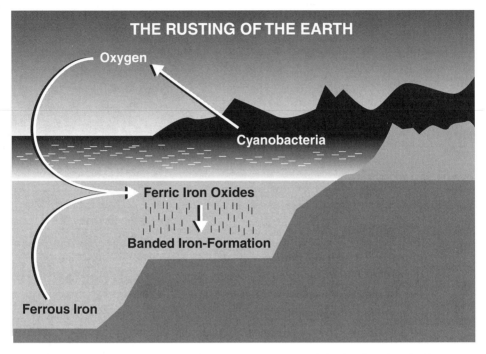

Figure 6.2 Earth rusted as iron from the deep ocean oxidized to form banded iron formations.

than 3,500 Ma to about 2,000 Ma ago, and are mined for steel worldwide, in Australia, Africa, Asia, Europe, North and South America. They are testament to a truly striking stage in planetary history—the rusting of the Earth—which drew to a close only when the ocean was finally swept free of dissolved iron. Yet the presence of these remarkable deposits does not mean the oceans were oxygen rich. On the contrary, BIFs were nearly always deposited in large basins, hundreds of kilometers in length and breadth, and the dissolved ferrous iron from which BIFs form could be spread over such expanses only if carried by waters that were oxygen poor. Huge amounts of molecular oxygen were pumped into the environment by oxygenic (cyanobacterial) photosynthesis, but except locally, near where it was produced, amounts of oxygen were kept low by its capture and rapid burial in the oxide minerals of BIFs.

Oxygen Levels Were Slow to Rise

That the world's environment contained only traces of oxygen up to roughly 2,000 Ma ago is shown also by other geologic indicators. The presence of O_2 strongly affects how some minerals resist weathering. Good examples are uraninite (UO_2, energy source for nuclear power plants) and pyrite (FeS_2, fool's gold), minerals that in today's oxygen-rich surroundings dissolve and weather away quickly. Yet in terrains older than about 2,200 Ma they make up major ore bodies, impressively large conglomeratic deposits formed at the mouths of ancient rivers (see plate 6). The persistence of granules and pebbles of uraninite and pyrite during weathering, transport in streams and rivers, and burial in thick deltaic conglomerates could not have taken place had much oxygen been present. The scarcity of oxygen earlier than 2,200 Ma ago is shown also by the chemistry of ancient soil horizons (paleosols), which reveals that the amount of O_2 in the atmosphere first began to increase at about that time.

BIFs, uranium-rich pyritic conglomerates, ancient soils, and other geologic indicators confirm a dramatic worldwide shift in the environment between about 2,200 and 1,900 Ma ago. The amount of oxygen earlier was apparently less than 1% that in the atmosphere today, but increased to about 15% of the present level as global rusting came to an end with deposition of the last abundant BIFs.

Evidence both from fossils and from geology points to the existence of oxygen-producing photosynthesizers, cyanobacteria, by 3,500 Ma ago. The third horse of the paleobiologic troika, isotopic geochemistry, provides yet more proof of early photosynthesis, and like the varied kinds of fossils in the Apex chert shows that cyanobacteria were not the only microbes present.

Isotopic Evidence of Ancient Metabolisms

Chemical elements are the fundamental substances that make up matter. There are more than one hundred different kinds, including C, H, O, N, S, P, each consisting of atoms of a particular type. But atoms of an element can come in more than one variety. These are isotopes, forms of an element that differ slightly in weight but have almost the same chemical behavior.

Some isotopes decay radioactively, others are immutable. Carbon, like many chemical elements, has isotopes of both types. The isotope that decays weighs 14 mass units and for this reason is known as carbon-14 or simply ^{14}C. By spontaneously casting off from its interior a subatomic particle (in the form of a beta ray), ^{14}C transforms to an isotope of nitrogen, ^{14}N, which has the same weight but a stable atomic core. Radioactive carbon is converted to ^{14}N at a steady rate, so the amount yet unchanged in a fossil bone or tooth is an index of its age. This is the basis of the famous carbon-14 method of dating prehistoric human remnants. But because ^{14}C decays rapidly, only fossils aged less than 60,000 or so years can be dated by this technique. Radioactive isotopes of some elements last much longer, and though several of these (potassium, rubidium, samarium, uranium, thorium) are used to date minerals in rocks, none is plentiful enough in fossils to be used to tell their age.

In addition to radioactive ^{14}C, carbon has two immutable isotopes, ^{12}C and the slightly heavier ^{13}C. Preserved in sedimentary rocks, these and the stable isotopes of H, O, N, and S provide powerful clues to the metabolism of early life.

Cyanobacterial and Bacterial Photosynthesis

Because heterotrophs obtain their nourishment from autotrophs, even if indirectly through the food chain, autotrophs govern the isotopic composition of the living world. And the isotopic makeup of autotrophs is controlled largely by the mix of isotopes in the gases and simple nutrients they take in from the environment. For the stable isotopes of carbon this is a mix of two types of carbon dioxide, $^{12}CO_2$ and $^{13}CO_2$.

In the first step of photosynthesis, a matchmaking enzyme (usually one known as RUBISCO, abbreviated from its formal name, ribulose bisphosphate carboxylase/oxygenase) captures CO_2 and brings it together with hydrogen to produce a CHO-containing molecule (commonly, phosphoglyceric acid) that by further steps is converted to glucose. But $^{12}CO_2$ and $^{13}CO_2$ have different weights, so they are ensnared by RUBISCO-type enzymes with different ease. Molecules of the two types of carbon dioxide bounce about in the cell sap like tiny tennis balls—the lighter $^{12}CO_2$ quickly, the heavier $^{13}CO_2$ more slowly. Because the lighter molecules hit the enzyme more frequently, they are caught and bound more often into the first-formed product and, consequently, into glucose, the source of carbon for all other organic compounds of the organism. As a result of this enzyme-guided process (technically, "kinetic isotopic fractionation"), the mix of atoms in the organic matter of autotrophs contains more ^{12}C, and correspondingly less ^{13}C, than the mixture in the atmospheric CO_2.

Carbon dioxide is captured also in minerals. When CO_2 dissolves in seawater it forms a chemical known as bicarbonate (HCO_3^-) that combines with soluble calcium (Ca^{+2}) to make mineral grains of calcium carbonate ($CaCO_3$) which rain onto the ocean floor to form limestone. This process separates the stable isotopes of carbon, but the effect is opposite that of photosynthesis—the isotopic mix in limestone contains less ^{12}C, and more ^{13}C, than the mixture in the CO_2 of the atmosphere.

Compared with atmospheric CO_2, photosynthetic microbes (like all photoautotrophs) are enriched in ^{12}C and limestones are depleted. The differences are small, a few percent or less, and by mass spectrometry are measured in number of parts per thousand (permil, ‰). So that they can be compared, measurements in laboratories worldwide are calibrated by use of a standard specimen (the PDB limestone), which has an assigned value of 0.0‰. Differences from this standard are expressed as $\delta^{13}C_{PDB}$ values (in which the Greek letter δ, delta, means difference) that for samples containing relatively more of the heavier isotope, ^{13}C, are positive, and for those having more ^{12}C are negative.

The divvying up of carbon isotopes that happens during photosynthesis leaves a telltale fingerprint in the rock record that can be used to track life's history. Compared to the CO_2 carbon source, photosynthetic microbes are enriched in the lighter isotope, generally by about 18‰, and limestones are depleted by 7‰. The two diverge by about 25‰, a clear-cut difference that can be traced into the distant geologic past: the $\delta^{13}C_{PDB}$ values of fossil organic matter cluster

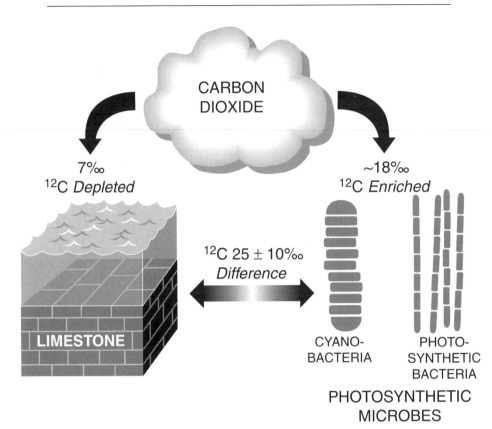

Figure 6.3 Microbial photoautotrophy partitions carbon isotopes between limestone and photosynthetic bacteria and cyanobacteria.

near $-25\%_0$ and those of ancient limestones around $0\%_0$, an imprint of photosynthesis that extends to 3.5 billion ago as, for example, in the Apex chert, where the difference is $27\%_0$.

The $\delta^{13}C$ values of ancient organic carbon vary considerably, by $\pm\ 10\%_0$, a range much broader than that of limestones. This difference in uniformity arises because different kinds of photoautotrophs separate carbon isotopes to different degrees, and these are affected by shifting environmental factors, whereas conditions of limestone formation have held more or less constant over time. Laboratory experiments show that cyanobacteria usually are a few to several permil less enriched in ^{12}C than photosynthetic bacteria in the same setting. In the Phanerozoic, this pattern can be traced to about 50 Ma ago in

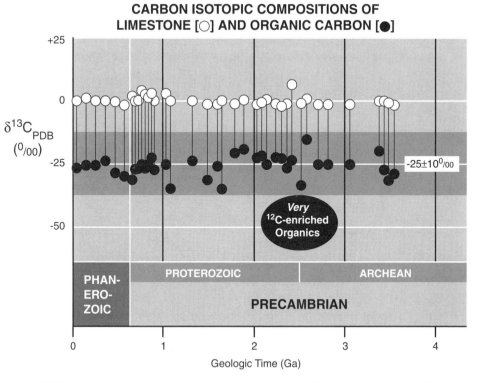

Figure 6.4 The carbon isotopic fingerprint of photosynthesis extends to 3.5 Ga ago.

fossil remnants of photosynthetic pigments, but the isotopic difference is blurred and can't be recognized in older deposits such as those of the Precambrian where organic remains have been aged and altered by heat and pressure.

Like the evidence from cellularly preserved fossils, carbon isotopes show that photosynthetic members of the Bacterial domain, probably cyanobacteria and photosynthetic bacteria both, have existed since 3,500 Ma ago. And measurements on graphitic carbon from the 3.8 billion-year-old Isua sequence hint that photosynthesizers may have been present even then.

Bacterial Sulfate-Reduction

Bacteria are either Gram-positive or Gram-negative, grouped by whether their encasing walls and membranes hold color when treated with

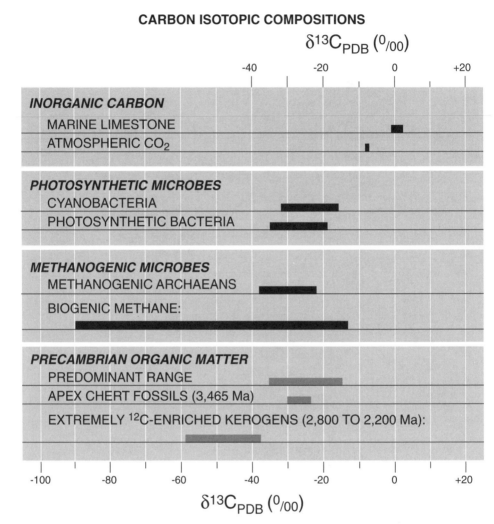

CARBON ISOTOPIC COMPOSITIONS

$\delta^{13}C_{PDB}$ (⁰/₀₀)

INORGANIC CARBON
  MARINE LIMESTONE
  ATMOSPHERIC CO_2

PHOTOSYNTHETIC MICROBES
  CYANOBACTERIA
  PHOTOSYNTHETIC BACTERIA

METHANOGENIC MICROBES
  METHANOGENIC ARCHAEANS
  BIOGENIC METHANE:

PRECAMBRIAN ORGANIC MATTER
  PREDOMINANT RANGE
  APEX CHERT FOSSILS (3,465 Ma)
  EXTREMELY ¹²C-ENRICHED KEROGENS (2,800 TO 2,200 Ma):

$\delta^{13}C_{PDB}$ (⁰/₀₀)

Figure 6.5 Inorganic carbon, living microbes, and Precambrian organic matter have telltale carbon isotopic compositions.

Gram's stain, a set of dyes developed in 1884 by the Danish physician Hans Christian Gram. The Universal Tree of Life shows that both kinds evolved early, but the nonstainable (Gram-negative) group, to which cyanobacteria and photosynthetic bacteria both belong, is especially well represented in the Precambrian rock record. The geochemistry of sulfur isotopes shows that a third kind of Gram-negative microbe, chemoautotrophs called "sulfate-reducing bacteria," dates also from early in Earth history.

Sulfate-reducing bacteria make energy by uniting hydrogen with sulfur atoms cut away from the dissolved sulfate (SO_4^-) of seawater to form hydrogen sulfide (H_2S), the vile-smelling gas of rotting eggs ($4H_2 + H_2SO_4 \rightarrow H_2S + 4H_2O + 39$ kilocalories). As in photosynthesis, this process separates two stable isotopes, but of sulfur rather than carbon, so that the gas given off is enriched in the lighter isotope, ^{32}S, relative to the mix of ^{32}S and ^{34}S in its sulfate source. The H_2S then combines with iron in sediments to form brassy grains of pyrite, fool's gold, and because the gas is enriched in the lighter isotope, the fool's gold is too.

Compared to the mix of ^{32}S and ^{34}S in the sulfur isotope standard (the iron-sulfur mineral troilite from the Cañon Diablo meteorite), the dissolved sulfate, which on evaporation forms sulfate minerals such as gypsum ($CaSO_4 \cdot 2H_2O$), is depleted in ^{32}S by about 20‰. Because the H_2S given off by sulfate-reducers is enriched by about 20‰, the isotopic difference between evaporitic sulfate and bacterially formed pyrite is generally about 40‰. But as in photosynthesis, the amount of isotopic fractionation that takes place when these chemoautotrophs convert sulfate to hydrogen sulfide varies as the environment changes. So in addition to being enriched in ^{32}S, microbe-produced pyrite ranges broadly in isotopic makeup and is easy to distinguish from pyrite grains weathered out of basalts, granites, and other igneous rocks that have tightly clustered compositions. These indicators show that anaerobic sulfate-reducing bacteria played a prominent role locally, and were represented globally, early in Earth history. Sulfur isotopes trace their existence to 2,700 Ma ago, and with less certainty to 3,400 Ma ago.

Archaeal Methanogenesis

Other isotopic studies show that the Archaeal domain also has ancient roots. Among the earliest evolved archaeans are those that make energy and give off methane by combining hydrogen with carbon atoms from carbon monoxide ($2H_2 + 4CO \rightarrow CH_4 + 3CO_2 + 50$ kilocalories), carbon dioxide ($4H_2 + CO_2 \rightarrow CH_4 + 2H_2O + 32$ kilocalories), or carbon stripped away from small organic compounds.

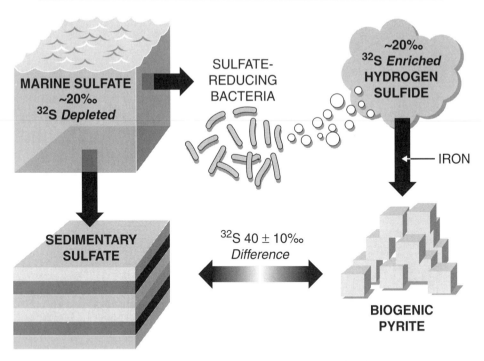

Figure 6.6 Bacterial sulfate-reduction results in partitioning of sulfur isotopes between sedimentary sulfate and biologically produced pyrite.

Like photoautotrophy, this chemoautotrophic lifestyle results in ^{12}C enrichment of growing cells, typically by about 30‰ relative to the PDB standard.

Because the $\delta^{13}C_{PDB}$ values of the cells of methane-producing archaeans overlap with those of photosynthetic microbes, they cannot be used to cleanly tell them apart. But the methane that archaeans exhale carries an identifying isotopic signature. Archaeal methane can be extraordinarily enriched in ^{12}C, by up to 90‰ relative to PDB, much more than any other biologically produced substance. Because the maximum enrichment attainable by photosynthesis is less, only 36‰, ^{12}C-rich methane leaves an unmistakable sign in the rock record. In the form of organic matter having $\delta^{13}C_{PDB}$ values of -38 to $-59‰$, carbonaceous remnants of microbes that used archaean-generated methane as their source of carbon, this telltale signature is displayed in more than a dozen widely scattered geologic units 2,800 to

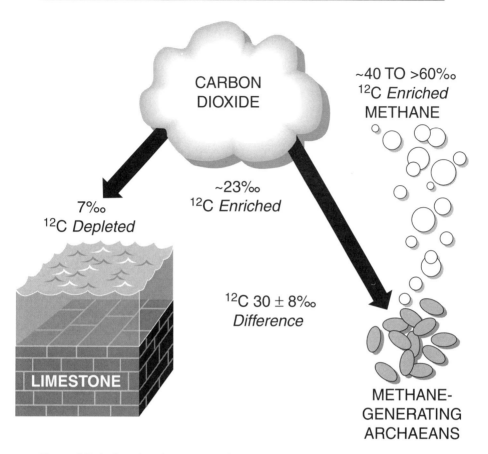

Figure 6.7 Archaeal methane-generation results in partitioning of carbon isotopes between limestone, methanogenic archaeans, and the methane given off.

2,200 Ma old. The wide geographic spread of these strictly anaerobic, chemoautotrophic archaeans fits well with the geologic evidence of an oxygen-poor environment at this time.

Paleobiology: Direct Evidence of Early Evolution

Fossils, minerals, and signals from isotopes all testify to life's great antiquity. Only a billion or so years after the planet formed, its surface swarmed with a microbial zoo of eaters and eatees, heterotrophs and both photo- and chemoautotrophs, oxygen producers and consumers, microbes needing no oxygen at all and those spewing forth hydrogen sulfide and methane gases. As Darwin saw more than a century ago, the only direct proof of life's long history is written in the rocks of the Earth. Read by fossils, geology, and isotopic geochemistry, together— the new science of Precambrian paleobiology—this record reveals that life evolved far and fast astonishingly early.

Stromatolites: Earth's First High-Rise Condos

Nature Is Not Compartmentalized

Though first noted by geologists in the early 1800s, the lumpy, layered, *Cryptozoon*-type rock masses we now know as stromatolites remained controversial until the 1960s. As we saw in chapter 1, some workers argued for a biologic origin while others, notably A. C. Seward and his followers, claimed they were purely inorganic. No one knew for sure until 1961, when word began to spread that living stromatolites had been discovered at a remote super-salty lagoon called Shark Bay about 1,000 kilometers north of Perth on Australia's desolate west coast.

Uncertainty about the true nature of stromatolites should never have lasted so long. The examples at Shark Bay are hard and rocky, so the geologists who first found them more or less instantly saw their link to the enigmatic *Cryptozoon*. But structures almost the same—soft or leathery, instead of rock solid, and called microbial mats instead of stromatolites—were already known to microbiologists who years earlier had shown their biologic origin. Yet because these studies had gone unnoticed by the geologic community, how stromatolites are built remained a mystery to geologists long after microbiologists figured it out!

It's fair to ask why the findings of one school of science didn't carry over to another. The answer is as simple as it is dismaying: microbiologists are trained in the life sciences, geologists in the physical sciences, and the two tribes have such different backgrounds and interests they barely speak the same language. On most college campuses they even occupy separate "homelands," and though their shared borders are said to be open they actually are seldom crossed, only occasionally by foraging students, even more rarely by faculty. Each tribe stresses its own specialized brand of knowledge, and as each prods its students to learn more and more about less and less, science becomes increasingly fragmented. There's little doubt this tribalism makes things simpler for all—learning the ropes in a single subject is far easier than grappling with many. But it also costs, and the price is high for those studying the real world, which, by its nature is a dynamic interlocking mix of the life *and* physical sciences. Certainly, if microbiologists and geologists had paid more heed to each other's findings we would not have had to wait so long for the *Cryptozoon* controversy to be resolved.

Stromatolites: Earth's First High-Rise Condos

What Are Stromatolites?

Like many terms in science, "stromatolite" is derived from ancient Greek, a combination of *stromatos*, "bed covering," "layer," and *lithos*, meaning "rock" (and the Greek root of such common geologic terms as lithology, the study of rocks; lithify, to convert to rock; and lithosphere, the outer rocky shell of the Earth). So, from the term itself we can guess that stromatolites are "layered rocks"—but whether they are rocky like those in the fossil record, or leathery to gelatinous like many living today, it is the biologic, microbial origin of the layering in stromatolites that makes them special.

The formal definition of these distinctive structures has four parts: a stromatolite is (1) an accretionary organosedimentary structure, (2) commonly thinly layered, megascopic, and calcareous, (3) produced by the activities of mat-building communities of mucilage-secreting microorganisms, (4) mainly filamentous photoautotrophic prokaryotes such as cyanobacteria.

Let's break this apart piece by piece to see what it means. The first part says that a stromatolite is an *accretionary organosedimentary structure*. "Accretionary" means that the structure builds up (accretes) layer by layer, and "organosedimentary" means that it forms through interaction of biologic (organo-) and physical (sedimentary) processes. In the second part, the structures are described as usually *thinly layered* (made up of stacked laminae, each often thinner than onionskin), *megascopic* (large enough to see without a microscope or even a magnifying lens), and *calcareous* (in part composed of calcium carbonate minerals such as calcite, the chalk-white grains that make up limestone).

The third phrase tells us a stromatolite is formed *by mat-building communities of mucilage-secreting microorganisms*. The term "mat" refers to any of the stacked organic-rich stromatolitic layers that in living examples, are usually tough, flexible, matlike sheets; "community" indicates that the mats are built by several or many kinds of microorganisms that live together in a natural biologic assemblage; and "mucilage-secreting" tells us these various organisms give off from their cells sticky gelatinlike slime (chiefly various carbohydrates that, clumped together with the living cells, make up the matlike sheets). The final phrase indicates that the mat builders generally are *filamentous* (rather than ball-shaped, for instance), *photoautotrophic* (photosynthetic "light-using self-feeders"), *prokaryotic* (bacterial and archaeal microbes, rather than eukaryotes), *and often include cyanobacteria* (the predominant organisms in the uppermost mat, the growth surface, of most stromatolites).

This definition, like most, contains hedge words. It's true that stromatolites *commonly* are thinly layered, megascopic, and calcareous— but some have layers several millimeters thick; some are microscopic, only tens of microns high; and rather than being calcareous, living stromatolites are often made mostly of organic matter instead of mineral, and some fossil ones are made of phosphate minerals, others of chert, and even the calcareous ones often have cherty parts (in most, chemically precipitated silica that takes the place of carbonate minerals originally laid down). And it's true also that the mat-building microorganisms are *mainly* filamentous photoautotrophic prokaryotes including many kinds of cyanobacteria. But spheroidal microorganisms, especially colony formers, also build stromatolitic layers. Prokaryotes normally dominate, but eukaryotes are present in some modern stromatolites (such as those at Shark Bay which periodically are infested by salt-tolerant diatoms). And rather than being formed by cyanobacteria, the layers just below the cyanobacterial growth

surface are often made by non-oxygen-producing photosynthetic bacteria—accomplished mat builders that early in Earth history, before
cyanobacteria orginated, might well have constructed stromatolites on
their own, as they do in rare environments today (showing that in and
of themselves, stromatolites should not be taken as proof positive of
cyanobacteria or O_2-producing photosynthesis).

Clearly, stromatolties are fairly complicated, so it's not surprising
they have been defined by various workers in various ways. Rather
than highlight the link between living and fossil examples, as the
definition used here does, some prefer to restrict "stromatolite" to
geologic specimens and use "microbial mat" for their living counterparts. Others call structures stromatolites only if they form sizable
bumps (that is, have fairly "high relief" above the neighboring
sediment) and refer to the flat-lying ones as mats or sheets. The particular definition doesn't matter too much as long as its meaning is
made clear and it stresses, as in the definition used here, that *it is the
biologic origin of the layering in stromatolites that makes them distinctive*. But the practice adopted by some of dubbing almost any
thinly layered calcareous rock "stromatolitic" should be avoided since
it confuses true stromatolites with nonbiologically deposited look-
alikes—cave rocks (such as stalactites and stalagmites) and some hot
spring deposits, for instance, formed where minerals build up in thin,
sometimes wavy layers as they crystallize from solution.

Mat-Building Microbial Communities

Biologically, a stromatolite is a thriving, active, microbial zoo, a self-
contained, small-scale ecosystem of anaerobes and aerobes, eaters and
eatees. To see how the zoo operates and why it's structured the way it
is, we'll first explore the microbial makeup of living stromatolites.
We'll then be ready to examine fossil examples and ask how similar
they are to mat-building communities living today.

The key players in modern stromatolites are cyanobacteria. As seen
in earlier chapters, they are a remarkable group of microbes, ancient
inventors of the oxygen-generating type of photosynthesis that drives
the modern ecosystem. Though cyanobacteria are advanced members
of the Bacterial domain, they nevertheless are rather simple—all are
tiny, nonnucleated, and with few exceptions composed of monoto-

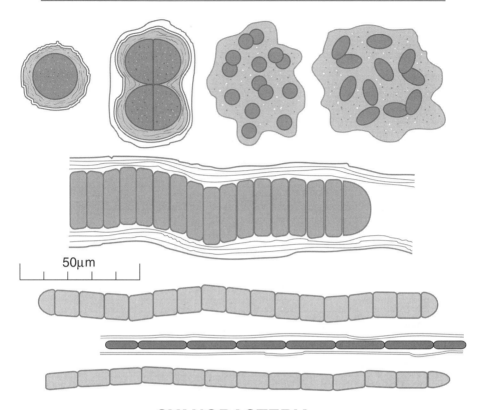

CYANOBACTERIA

COCCOIDAL, ELLIPSOIDAL, AND FILAMENTOUS

Figure 7.1 Cyanobacteria come in a variety of shapes and sizes and are often surrounded by or embedded in secreted layers of sticky mucilage.

nously unvarying cells able to both produce oxygen by photosynthesis and consume it by aerobic respiration. But they are usually larger than other prokaryotes and have a much broader range of forms, the most common being ball shaped (coccoidal), egg shaped (ellipsoidal), and stringlike (filamentous). They are key to the way stromatolites form because they make up the dense microbial mat at the uppermost surface where the structures grow by adding new stacked layers one by one.

Three traits account for why cyanobacteria form mats. First, to photosynthesize they need to see the Sun, a requirement they meet by spreading across surfaces as a living, sheetlike film. Second, almost all are embedded in sticky slime secreted from their cells—sometimes in copious amounts, other times not, depending on changes in the local environment—and this slime provides the glue that holds a mat together. Third, the filamentous kinds are especially good at building mats because their long chains of cells can glide—toward light, away from light, depending on how intense it is (a behavior called "phototaxis"). Their top speed seems unimpressive, only a few millimeters per hour, but for micron-sized microbes it actually is quite respectable (and scaled to human terms equates to a pace between an amble and a fast walk). Because they can glide, the filaments intermingle and become tangled and entwined as they battle for sunlight, and this gives the mats an interwoven feltlike fabric that explains why they are flexible and robust.

In almost all living mat-building communities, just beneath the cyanobacterial mat is a thin zone, the undermat inhabited by photosynthetic bacteria, most commonly kinds known from their pigmentation as green sulfur and purple bacteria. The undermat is umbrellaed by a dense overlying layer of cyanobacteria, so it seems puzzling that it is made up of microbes that need light to grow. However, this presents no problem to the photosynthetic bacteria since they can actually "see" through the cyanobacteria above them. Like all oxygenic photoautotrophs, cyanobacteria use chlorophyll-*a* to capture the light energy that powers their photosynthesis. But the anoxygenic photosynthesizers of the undermat instead use bacteriochlorophylls and other pigments, and these absorb wavelengths of light that pass through cyanobacteria. As we saw in chapter 5, this neat scheme arose early in Earth history when photosynthetic bacteria and their mutant offspring, the first cyanobacteria, competed for photosynthetic space, and it explains why the layering in living microbial mat communities is often rainbowed (see plate 7).

Oxygen is plentiful only in the cyanobacterium-dominated growth surface where it is produced, so this is the only stromatolitic zone inhabited by microbes that cannot survive without oxygen (technically, "obligate aerobes"). Most of the oxygen is used either by the cyanobacteria themselves or by cohabiting aerobic bacteria, and whatever is left usually diffuses to the atmosphere. Practically no oxygen

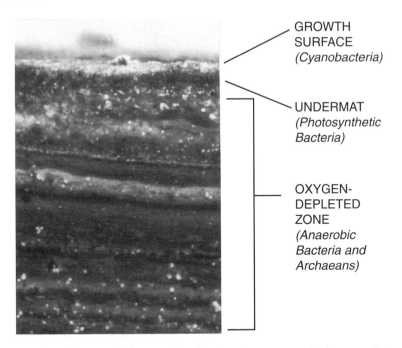

GROWTH
SURFACE
(Cyanobacteria)

UNDERMAT
*(Photosynthetic
Bacteria)*

OXYGEN-
DEPLETED
ZONE
*(Anaerobic
Bacteria and
Archaeans)*

Figure 7.2 As shown in this vertically sliced modern stromatolite from a salty lagoon in Baja, Mexico, mat-forming microbial communities are usually composed of three distinct zones—an uppermost bluish-green layer, the growth surface, made up of cyanobacteria; a thin smoky green and purple undermat layer of photosynthetic bacteria; and a thick, lowermost, oxygen-depleted zone inhabited by anaerobic bacteria and archaeans. (Bar for scale represents 1 cm. Color photographs of this specimen are shown in plate 7 A, B.)

seeps down into the undermat, but when it does it is used instantly by bacteria called "facultative aerobes" because they are able to consume oxygen (via aerobic respiration) if it's present but switch to fermentation (anaerobic metabolism) if it's not. These switch-hitting facultative forms keep oxygen levels in the undermat near zero, and in this way protect the photosynthetic bacteria that are poisoned by even trace amounts of free oxygen. Beneath the undermat is a sometimes centimeters-thick oxygen-depleted zone inhabited by bacteria and archaeans that cannot grow in the presence of oxygen (and are therefore obligate anaerobes), often including the chemoautotrophic sulfate-reducing and methane-generating kinds that as we saw in chapter 6, leave telltale isotopic signatures in the geologic record. The upper parts, and sometimes even the deep parts, of the oxygen-depleted

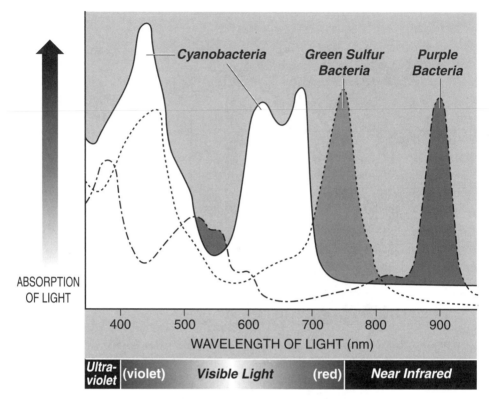

Figure 7.3 Despite living in the undermat where they are shaded by the cyanobacterial growth surface, green sulfur and purple bacteria are able to carry out photosynthesis because their pigments harvest red and infrared wavelengths of light not absorbed by the overlying cyanobacteria.

zone are usually layered and can be rather brightly colored (see plate 7), but instead of being produced by anaerobes living in the zone these are relicts left from when the buried layers and pigments were made by photosynthesizers in the growth surface and undermat, and they fade away over time as the pigments break down and anaerobes devour the CHON in the buried mats.

How Do Stromatolites Grow?

We can now see that mat communities are bustling menageries of microbes zoned top to bottom by the two necessities of life, CHON and energy. Photosynthesizers capture light energy and use it to build

CHON-rich organic matter at and near the growth surface. The organic compounds that are built are ultimately turned to fodder, since when the photosynthesizers die the compounds filter through the food web to the masses of heterotrophs below as each extracts sustenance from whatever foodstuff trickles down to its level in the pile or is provided by nearby chemoautotrophs. Some like to think of stromatolites as Earth's first high-rise condos, and in many ways they are. But they're probably even more like old-style New York City apartment houses turned upside down, with ground-floor mom-and-pop grocery stores (photoautotrophs) moved to the penthouse suites where they provide staples for the apartment dwellers below. On lower floors the fodder they provide is supplemented by scattered produce-stocked vending machines (chemoautotrophs).

Though we now know why food-making cyanobacteria occupy the penthouse, we haven't yet seen why the apartment houses have so many floors. What causes stromatolites to build up layer by layer? Again the answer comes from modern microbial mats. The growth surface of a stromatolite can be one of many shapes—flat, wrinkled, bumpy, domical, conical—depending on the organismal architects and the environment where it forms, and like all layers in a stromatolite, the growth surface tends to follow closely the contours of the layer just below it. When conditions remain constant, the growth surface doesn't change—aged cyanobacteria die, young ones take their place, the mat retains its shape, new layers are not added.

But when conditions do change—for instance, after a spring rain floods the growth surface with mud—the cyanobacteria respond. The prime directive for all of life is to stay alive, and to do this cyanobacteria must have sunlight. So, if a muddy layer blocks out the Sun's rays, the filamentous kinds respond by sloughing off their mucilage investments and gliding upward through the accumulated silt to find the new sunny surface, which they then rapidly colonize. Since the photosynthetic bacteria in the undermat need to harvest sunlight too, if the cyanobacteria move upward the undermat microbes follow suit, and this in turn frees space soon occupied by anaerobes that move up from below to feed on whatever organic matter has been left behind. Because construction of a new growth surface is usually a response to burial of the old, and because burial happens sporadically (whether by

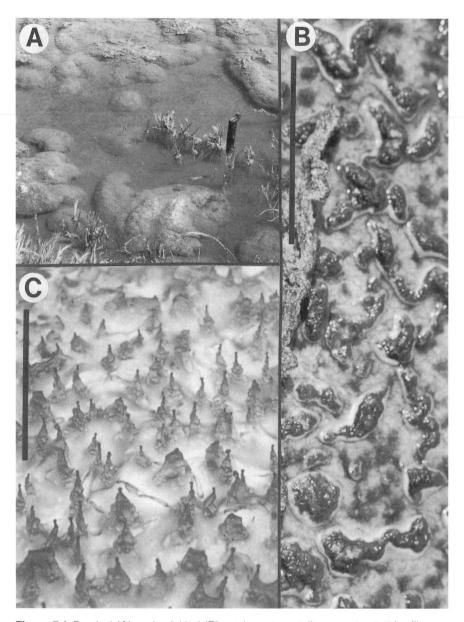

Figure 7.4 Domical (**A**) and wrinkled (**B**) modern stromatolites constructed by filamentous cyanobacteria at a salt marsh in Baja, Mexico. (**C**) Conical stromatolites built by the filamentous cyanobacterium *Phormidium* in a hot spring at Yellowstone National Park, Wyoming. [Scale of (A) shown by machete at right center; bars for scale in (B) and (C) represent 10 cm. (C) courtesy of M. R. Walter, Macquarie University, Australia.]

flooding muds, storm debris, wave-washed sediment, or precipitated mineral grains), the layers in stromatolites cannot be used like tree rings to tell a stromatolite's age. In some settings stromatolites accrete rapidly, but as a rule they build very slowly, a millimeter or less over several or many years.

Though some living microbial mats are tough and rubbery, most have the consistency of dried-out cream cheese and are soft enough to be sliced with a pocket knife (see plate 7). This seems at odds with the abundance of stromatolites in the Precambrian rock record, for it is hard to see how something so soft could readily be fossilized. The answer, of course, is that the soft ones almost never get preserved. Occasionally, soft mats become petrified (which when it happens rapidly, before decay sets in, can give beautifully preserved fossil cells), but the great majority of soft gelatinous mats are squashed, flattened, and "fubarized" in the rock-forming process. Rocky stromatolites, however, *are* easily preserved, especially if, like those at Shark Bay, they are lithified by grains of limestone-forming calcite (or its mineralic precursor, aragonite) precipitated onto the growth surface while the stromatolite-building microbes are still alive.

Modern Rocky Stromatolites

The living stromatolites found in the early 1960s at Shark Bay on Australia's west coast were a real boon to geologists since they are rock solid, just like the fossil ones. Other lithified examples have been found since, in the Bahamas and elsewhere, and all have rough rocky growth surfaces that mask insides made up of stacked layers formed the same way as those in soft microbial mats. The ones at Shark Bay have been studied most and are of special interest since in some ways they bear an uncanny resemblance to stromatolites of the Precambrian.

But before we turn to the fossil versions, there's one more question we need to answer, namely, why are rocky living stromatolites so rare? They're known from Shark Bay, the Bahamas, and a few other places, but in the Precambrian they populated the globe, persisting for billions of years along the shores of ancient seas and in lakes and ponds, rivers and streams. Why don't we find them there today? The answer is easy: life evolved and the cast of players changed. To

Figure 7.5 The rocky columnar stromatolites at Shark Bay on Australia's west coast have a rough, gnarly growth surface (**A**) that masks an ordered, many-layered interior (**B**) made up of a stacked sequence of earlier growth surfaces.

some organisms, cyanobacteria are a food of choice—snails especially seem to find them tasty. But snails and other many-celled grazing animals didn't exist until near the beginning of the Phanerozoic, about 550 Ma ago. Before then the grazers (protozoans, for instance) were not nearly so voracious, so rather than being devoured the slow-growing stromatolites had time to build. They can grow at Shark Bay only because it is a hypersaline lagoon, too salty for the eaters to tolerate; and microbial mats in hot springs, super-salty ponds, and along scorching desert coasts are able to form for the same reason: if conditions are too harsh for the eaters, the eatees can flourish.

Stromatolites of the Geologic Past

Are Today's Stromatolites Like Those of the Precambrian?

Stromatolites have been known to geology for well over a century during which various workers at various times attempted to sort the structures into manageable groupings. Before living stromatolites were found and microfossils were discovered in ancient ones, some workers thought that stromatolites were fossils of whole organisms, probably of a kind of seaweed, so many were formally described and given official Latin or Greek names. Even though we now know that stromatolites are built by communities of microbes, rather than being fossils of single organisms, many of the "genus" names are still in use since they provide handy monikers familiar to most workers (though now the genus names are called "group" names to avoid confusing them with the names of real biologic genera). Most stromatolites can be classed in the four most common kinds: flat-layered (the group *Stratifera*), domical (like *Cryptozoon*), columnar (groups such as *Gymnosolen* and *Colonnella*), and conical (*Conophyton*).

The shape of a stromatolite is governed mainly by the setting in which it builds. The flat-layered ones tell us the environment was quiet, perhaps a shallow lagoon or stagnant salty marsh. The domical and columnar kinds grade one into the other, as they do in the tide-washed waters at Shark Bay; they show that the setting was in places turbulent enough to rip up mats that connected one stromatolite to the next (producing isolated domes) but in other places was more quiescent (allowing linking mats to form between neighboring columns).

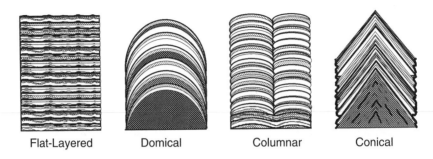

| Flat-Layered | Domical | Columnar | Conical |

TYPES OF STROMATOLITES

Figure 7.6 Depending mainly on the setting in which they form, stromatolites can be flat-layered, domical, columnar, or conical.

The conical forms usually occupy deeper, subtidal waters where there is less light, and their shape may be due to the action cyanobacteria of the growth surface gliding to the highest bump on the mat, then piling up in a peaked mass as they jostle for photosynthetic space.

Like the rock record itself, the abundance of stromalite-bearing beds peters out in increasingly older terrains: more than a thousand date from the more recent part of the Proterozoic, hundreds from the older Proterozoic, and fewer than three dozen from the Archean. Only four or five stromatolitic horizons older than 3,200 Ma have been discovered, a scarcity that has prompted some workers to question whether the stromatolites described from these especially ancient deposits might actually owe their origin to physical, geologic processes—the slumping and folding of soft, watery sediment layers as they solidified to rock—rather than to stromatolite-building microbes. Though a purely nonbiologic origin cannot be ruled out for some of these structures (particularly, sharp-tipped forms that occur as rare, isolated, or scattered foldlike bodies), there seems to be no reason to doubt the microbial genesis of the flat-layered, gently domical, or columnar stromatolites of this age, most of which are practically identical to younger stromatolites unquestionably formed by microbes. Even certain of the conical varieties defy a nonbiologic explanation: for example, scores of specimens are in places packed together cheek by jowl in a horizon of the Western Australia Pilbara sequence (dated to be 3,450 Ma old and therefore arguably the oldest stromatolitic bed known) that extends over a reported "tens of square kilometers," a spread much too broad to be accounted for by localized soft-sediment deformation.

So, we know that the shape of a stromatolite is determined mostly by its environment and that all of the common stromatolite shapes extend into the distant geologic past. Evidently, since the range of environments on Earth has not changed much over geologic time, neither have the stromatolites. In overall appearance, flat-layered stromatolites changed little or not at all; the reefs of domical and columnar ones at Shark Bay closely resemble fossilized reefs in South Africa 2,300 Ma old, half the age of the planet; specimens of one-billion, two-billion, and more than three-billion-year-old columnar stromatolites all look nearly the same; and the conical kinds, less common than the others, seem not to have changed much, either. What this means is that stromatolites can tell us about past environments, but probably not very much about evolution. Early workers tried to use them as index fossils to tell the relative ages of the strata in which they are preserved. But this practice has been largely abandoned since they change little over time and are not fossils of individual organisms, so, of course, they can hardly be expected to have evolved in the way that index fossils of the Phanerozoic did.

Are Stromatolites Subject to the Volkswagen Syndrome?

Even though stromatolites didn't evolve, one would think the microorganisms that built them must have, giving rise to what can be called a "Volkswagen Syndrome"—a lack of change of external form that masks internal evolution of the working parts. This can be checked by examining the microscopic fossils petrified in ancient stromatolites, the cellular and tubular filaments and single-celled and colonial spheroids that make up their mats, and comparing these to microbes that build mats today. In chapter 2 we've already seen a couple of well-preserved microbial mat-building communities—those of 850-Ma-old flat-layered stromatolites of the central Australian Bitter Springs Formation and 2,100-Ma-old domical stromatolites of the southern Canadian Gunflint chert. In chapter 3 we were introduced to a third assemblage, likely stromatolitic too (though we can't know for sure until the source bed of its fossil-bearing pebbles is found), from the nearly

Figure 7.7 Domical and columnar stromatolites built today at Shark Bay, Western Australia (**A**), look almost identical to fossil examples billions of years old (**B**, from the 2,300-Ma-old Transvaal Dolomite of Cape Province, South Africa). [The geologic hammer in (A), at right center, shows the scale of both photos.]

Figure 7.8 Comparison of vertically sliced specimens shows that columnar stromatolitic mats changed little over geologic time. Mats 1,300 Ma old (**A**, from the Belt Supergroup of Montana) are practically indistinguishable from those 2,000 Ma old (**B**, from the Albanel Formation of Quebec, Canada), and both closely resemble mats dating from 3,350 Ma ago (**C**, from the Fig Tree Group of the eastern Transvaal, South Africa). [Bars for scale represent 1 cm. (B) courtesy of H. J. Hofmann, University of Montreal; (C) courtesy of D. R. Lowe, Stanford University.]

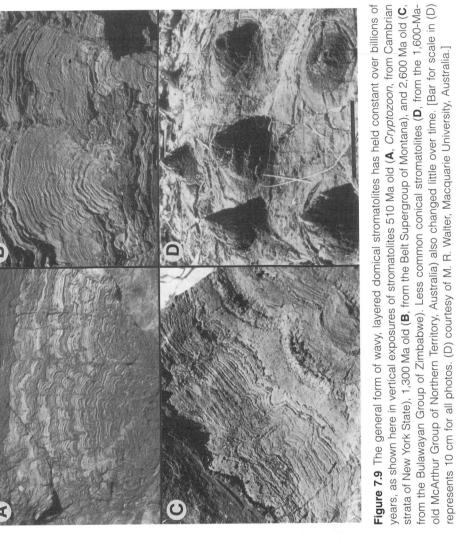

Figure 7.9 The general form of wavy, layered domical stromatolites has held constant over billions of years, as shown here in vertical exposures of stromatolites 510 Ma old (**A**, *Cryptozoon*, from Cambrian strata of New York State), 1,300 Ma old (**B**, from the Belt Supergroup of Montana), and 2,600 Ma old (**C**, from the Bulawayan Group of Zimbabwe). Less common conical stromatolites (**D**, from the 1,600-Ma-old McArthur Group of Northern Territory, Australia) also changed little over time. [Bar for scale in (D) represents 10 cm for all photos. (D) courtesy of M. R. Walter, Macquarie University, Australia.]

3,500-Ma-old Apex chert of northwestern Australia. But these three benchmark communities, known for their contributions to the development of the science and understanding of life's early history, are only a tiny sample of more than 250 that have been found in Precambrian stromatolites worldwide, a list of locales that reads like the United Nations—Australia, Brazil, Canada, China, France, India, Israel, Kazakhstan, Norway (Svalbard), Russia, South Africa, the USA.

So, there are now plenty of examples of fossil communities that one can compare with their living, mat-building counterparts, and—remarkably—like the stromatolites they build, the mat builders themselves evidently didn't evolve, either! This absence of change seems so contrary to the usual view of Darwinian evolution that it needs to be fully documented and explained before it can be accepted. We'll explore this surprising finding in the next chapter. Suffice it here to say that there is strong evidence that the same kinds of microorganisms living in the same kinds of environments built the same kinds of stromatolites over billions of years.

What Are Stromatolites Good For?

Though ancient and distinctive, abundant and widespread in the early rock record, and a storehouse of geologic and biologic information, stromatolites must seem arcane oddities to the uninitiated. Indeed, though stromatolites are occasionally mentioned in newspaper reports, and I've seen the term used in *Time, Newsweek*, and most recently (March 1998) *National Geographic*, most people have never heard of them. It's fair to wonder why we ought to know about them, and what, if anything, they're good for. Of course, the small ones make nice paperweights; the big ones, good doorstops; and all are great "dust collectors" (my wife's term for the specimens scattered about our house). But they have other, more unusual uses as well.

Years ago, in 1982, soon after President Nixon and Chairman Mao normalized U.S.-China relations, my wife and I had the pleasure of visiting Beijing as guests of the Chinese Academy of Sciences to do research at the Academy's Institute of Botany. One afternoon, she telephoned me at my lab to tell me that the work benches in her laboratory were made of stromatolites. Since she's a botanist, not a

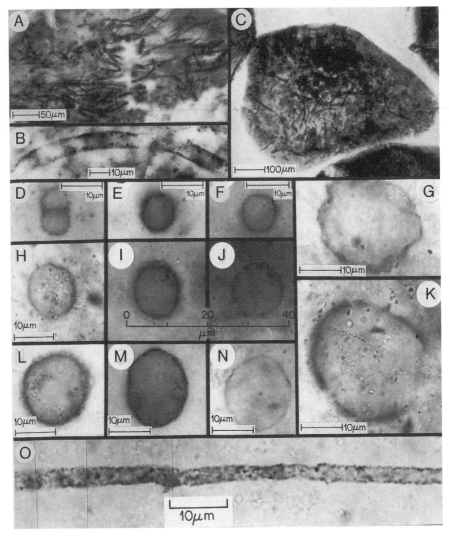

Figure 7.10 As in living stromatolites, the sheetlike layers in fossil stromatolitic communities are often made up of masses of intertwined cyanobacterial filaments (A and C), but single cells, cell pairs, and many-celled colonies are also common. (**A**) through (**C**) from the 1,950-Ma-old Tyler Formation of northern Michigan; (**D**) through (**O**), the 1,700-Ma-old Vempalle Formation of central India. (Other examples of microorganisms petrified in fossil stromatolites are shown in plates 1 and 2.)

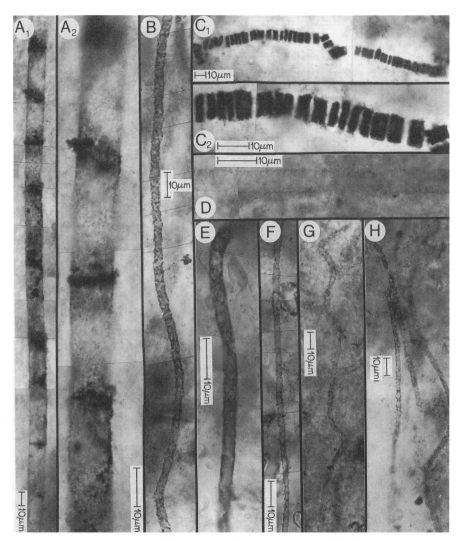

Figure 7.11 Stringlike cyanobacteria, some composed of long cells (A), others of short disk-shaped segments (C), are especially plentiful in petrified microbial mats, as shown by these examples from domical stromatolites of the 770-Ma-old Skillogalee Dolomite of South Australia. [Because the filaments are three dimensional and sinuous, most are shown in composite photos; (A) and (C) each show two views of a single specimen.]

geologist, I was a little skeptical, but when I hurried over to have a look I found she was absolutely right. All of her lab benches were built of limestone stromatolites (though unfortunately for me, the stromatolites were calcareous, not the cherty kind I sample when hunting for microfossils, so I left her benches unscathed). Moreover, weeks earlier, to the east of Beijing, I had studied the particular rock unit from which the limestone had come, so I knew the age of the stromatolites (1,500 Ma) and their source (near the village of Jixian). My wife and I both were pleased—it's fun to see scientific specimens put to practical use—so we decided to track down the company that built the benches to find out whether stromatolites were being used for anything else.

After a few phone calls we located the supplier—the People's Marble Factory and Work Unit in one of Beijing's suburbs—who, on learning of my interest in these curious structures, invited us to visit. A few days later we did. Though I'm not sure what I expected to find, I know I was impressed when I saw how extensive the factory was, made up of five huge, football-field-sized sheds. On our arrival we were escorted to the factory office, where we were greeted by the director and his deputy, and with my wife (a native-born Chinese, educated in the United States) acting as translator, we exchanged pleasantries over tea and Chinese cakes. Soon I was up at the blackboard in the small office explaining microbial mats, growth surfaces, cyanobacteria, microfossils. I don't know whether our hosts found this boring or not, but if they did they took pains not to let it show, and they definitely were amazed (and I suspect thought it a bit odd) to find that their "flower ring rocks," as they called them, were of such immense interest to a foreign academic.

When we finished our tea and cakes the director took us on a tour of the factory. Now it was my turn to be amazed, for the sheds were filled with stack after stack of cut and polished, beautifully prepared, museum-quality stromatolitic slabs—huge ones (3 meters square), small ones (about 10 centimeters across), and sizes in between—thousands of specimens! The variety was stunning (I toted up examples of five named stromatolite groups and several different subgroups) and spanned a range of colors—light gray with jet-black layers, dark gray set off by beige or light gray layers, dull red to pink and purple

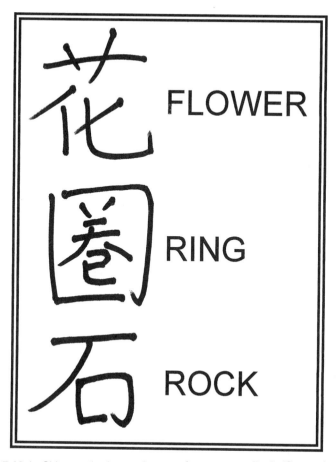

花　FLOWER

圈　RING

石　ROCK

Figure 7.12 In Chinese, the layman's name for a stromatolite is "flower ring (floral wreath) rock." Geologists use characters that translate to the more technical, but equally descriptive, "stacked layer rock." (Calligraphy by Jane Shen Schopf.)

flecked with whitish zones. This assortment was wonderfully impressive, yet we were told that this stromatolite factory wasn't even the biggest of nine scattered across China.

After the tour I wandered off along the back of the sheds to explore the waste piles of slabs broken during the sawing process, and there found the most spectacular stromatolite specimen I've ever seen—a large rectangular slab, missing only a sizable corner, of brilliantly bright-red columnar stromatolites (the group *Baicalia*) crossed by sharply defined veins of sparkling pure white calcite. The director soon found me hunkered over the pile of castoffs, and, seeing my interest in that particular slab, told me it had been cut and polished in 1976 to use as one of the facing stones on Mao Zedong's mausoleum in the heart of Beijing, but the order had been canceled at the last moment when someone realized that the slabs prepared were cut from soft limestone that would weather away in Beijing's acid rain. The slab with the broken corner was the only one they had left, which he offered, at a bargain rate, to size down to a smaller rectangle, re-polish, and ship to me in Los Angeles. Mao's slab (of China's 850-Ma-old Northeast Red Formation) arrived intact a few weeks later and is now embedded in the entryway to our home.

It gives me a lot of pleasure to have such a beautiful piece of Precambrian stromatolitic limestone in our hallway. And though it is a rarity in the United States, it would not be so in China, where stromatolites are familiar to many people, not just academics. They have been used in Chinese architecture for more than 2,000 years, so it is no surprise to find splendid examples in the Forbidden City (the emperor's palace grounds) or at Beijing's famed Temple of Heaven. Today, stromatolites are used in up-scale hotels to tile floors, walls, ceilings, bathrooms, reception desks; strikingly beautiful examples can be found making up the majestic columns that front the Great Hall of the People on Tiananmen Square; and at a children's park near the Beijing Zoo, I even found slabs of stromatolites used to construct a playground slide (see plate 8).

Pretty Stones That Say a Lot

But ancient stromatolites are more than merely useful pretty stones. They can answer crucial questions about the early environment and ancient life, but only if they are actually stromatolites, not non-biologic look-alikes. Fortunately, there are not many other kinds of rocks that have stromatolitelike layering, and these usually are fairly easy to tell apart from true stromatolites since they form when minerals repeatedly crystallize out of solution to make stacked surface-coating layers that are almost always more uniform, much thinner, and decidedly more regular than those laid down by life (and they of course never harbor fossil mat-building microbial communities). Once properly identified, stromatolites (1) tell us about the

Figure 7.13 Stromatolites in China's past. (**A**) Statue at Nanjing Teachers' College of "The Great Helmsman," Mao Zedong, on a pedestal of 850-Ma-old stromatolitic limestone (**B**, from the Northeast Red Formation of eastern China). (**C**) Ornate stairway (sculpted nearly two thousand years ago out of a stromatolitic slab of the 1,500-Ma-old Wumishan Formation quarried east of Beijing) over which the emperor was carried in a sedan chair during his yearly visit to Beijing's Temple of Heaven, where, from the center of a circular limestone stage (**D**), also made of Wumishan stromatolites, he sought heavenly blessings for a bountiful harvest.

environment where they formed, whether it was quiet or energetic, shallow or relatively deep; (2) reveal that a special kind of life was present—phototactic microorganisms that, to survive, needed to see the Sun; and (3) show that a full-fledged ecosystem existed made up of microbial food producers and consumers, nearly always including aerobes, facultative aerobes, and anaerobes. Moreover, as we'll see in the next chapter, stromatolites are a choice hunting ground for ancient cellular fossils and the source of much of our knowledge about the earliest history of life.

(Interestingly, none of these facts comes as news to NASA, including the especially serious problem of distinguishing true stromatolites from nonbiologic look-alikes. Microbes are so widespread on Earth that there is practically no place where the look-alikes can form without life playing a role. But on a planet where life never got started, there could be many places veneered by thinly layered stromatolite*like* deposits unrelated to life—laid down, for instance, by repeated wetting and drying or freezing and thawing of mineral-charged salt pans or shallow lagoons. So, though NASA's Mars rock-sampling mission won't lift off until the year 2005 and bring back rocks until three years later, NASA scientists are already working out ways to tell real stromatolites from "foolers" as they tool up to search the returned rocky debris for signs of martian life.)

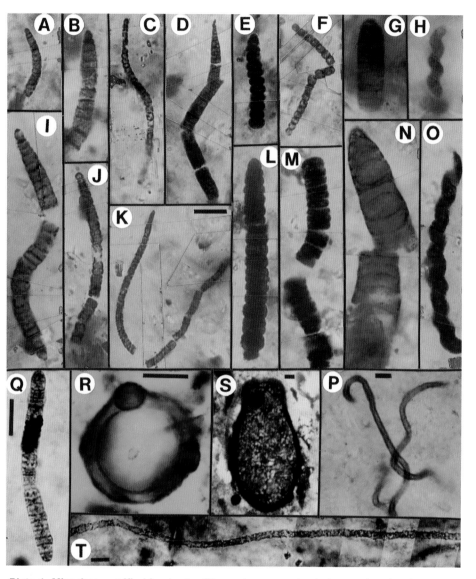

Plate 1. Microbes petrified in cherts. Filamentous cyanobacteria are abundant in stromatolitic cherts of the 850-Ma-old Bitter Springs Fm. (A–P), 1,500-Ma-old Barney Creek Fm.
(Q), and 770-Ma-old Skillogalee Dolomite (T) of Australia. Fossils rare in stromatolites include the double-layered spheroid in (R) from the 2,100-Ma-old Gunflint Fm. of Canada, and
the vase-shaped protozoan in (S) from the 850-Ma-old Kwagunt Fm. of Arizona. (**A, B, I,** and
J) *Cephalophytarion*; (**C** and **K**) *Cyanonema*; (**D**) *Caudiculophycus*; (**E** and **L**) *Filiconstrictosus*; (**F**) *Veteronostocale*; (**G**) *Palaeolyngbya*; (**H** and **O**) *Heliconema*; (**M** and **Q**) *Oscillatoriopsis*; (**N**) *Obconicophycus*; (**P**) *Eomycetopsis*; (**R**) *Eosphaera*; (**S**) *Melanocyrillium*; (**T**) unnamed trichome. [Because the petrified filaments are three dimensional and sinuous, most
are shown in composite photographs. Bars for scale represent 10 μm; the bar in (K) shows
the magnification of (A) through (O).]

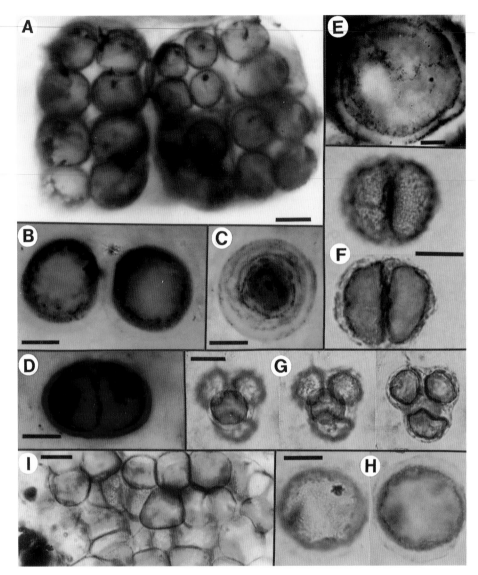

Plate 2. Spheroidal cyanobacteria in cherts. Petrified single and colonial unicells are plentiful in cherts of the 1,000-Ma-old Sukhaya Tunguska Fm. of Siberia (A–D), the 850-Ma-old Bitter Springs Fm. (F–H) and 770-ma-old Skillogalee Dolomite (E) of Australia, and the 650-Ma-old Chichkan Fm. of Kazakhstan (I). (**A**) *Eoentophysalis*; (**B**) unnamed chroococcaceans; (**C**) *Gloeodiniopsis*; (**D**) *Chroococcus*-like cell pair; (**E**) unnamed sheathed unicell; (**F**) *Eozygion*; (**G**) *Eotetrahedrion*; (**H**) *Globophycus*; (**I**) *Palaeopleurocapsa*. [(F), (G), and (H) show multiple views of single specimens. Bars for scale represent 10 μm. [(A) through (D) courtesy of C. V. Mendelson, Beloit College, Wisconsin.]

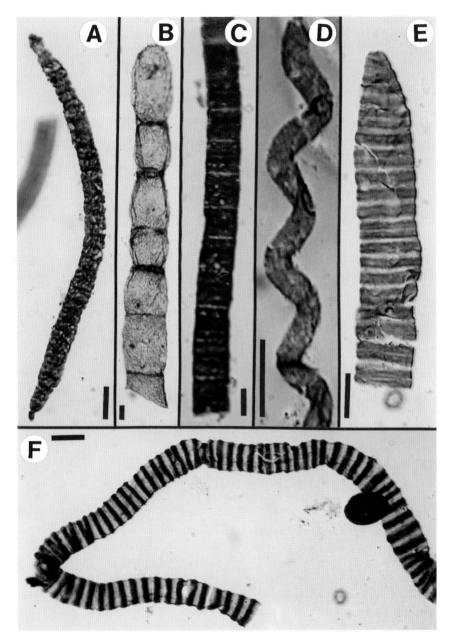

Plate 3. Flattened filaments in siltstones. Compressed Precambrian filaments are especially well known from shaley rocks of Siberia (A and B, from the 850-Ma-old Miroedikha Fm.; C, the 950-Ma-old Derevnaya Fm.) and Bashkiria (D, from the 925-Ma-old Sim Fm.; E and F, the 925-Ma-old Shtandin Fm.). (**A**) *Primorivularia*; (**B**) *Trachythrichoides*; (**C**) *Partitiofilum*; (**D**) *Heliconema*; (**E** and **F**) *Calyptothrix*. (Bars for scale represent 10 μm.)

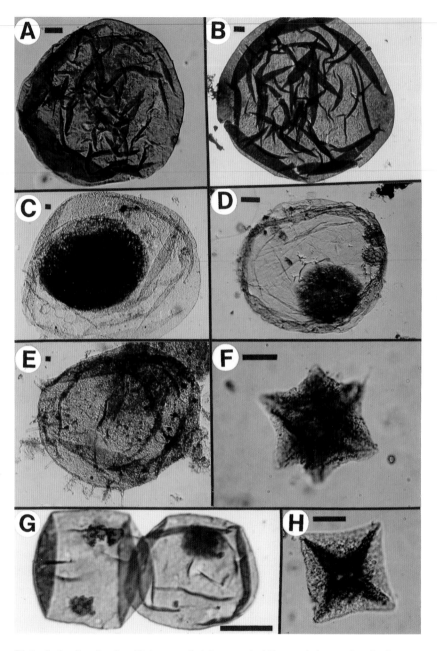

Plate 4. Acritarchs in siltstones. A rich record of Precambrian acritarchs is preserved in rocks of Siberia (A and F, from the 950-Ma-old Lakhanda Fm.; C, the 850-Ma-old Miroedikha Fm.), Bashkiria (D and G, from the 580-Ma-old Redkino Fm.), and the Ukraine (F and G, from the 600-Ma-old Derlo Fm.). (**A** and **B**) *Kildinella*; (**C**) *Pterospermopsimorpha*; (**D**) *Leiosphaeridia*; (**E**) *Trachyhystrichosphaera*; (**F** and **H**) *Octoedryxium*; (**G**) *Arctacellularia*. (Bars for scale represent 10 μm.)

Plate 5. Precambrian water and ice. Liquid water was present early in Earth history as shown by a pillow lava (**A**) from the Komati Fm. of South Africa, formed 3,500 Ma ago when volcanic rock solidified on contact with the sea, and by water-formed ripple marks (**B**) in the overlying 3,300-Ma-old Moodies Group. Precambrian ice is evidenced by the glacial pavement of the Smalfjord Tillite of Norway (**C**, courtesy of G. Vidal, University of Lund, Sweden), planed and grooved by a massive ice flow near the close of the Precambrian.

Plate 6. Precambrian oxygen. The rust-red iron oxide layers in banded iron formation (BIF) from the South African Hotazel Fm. (**A**) show O_2 was present in seawater when the rock formed 2,200 Ma ago. Sponged up by BIFs, free oxygen was scarce before about 2,000 Ma ago, as shown by a hand-sized specimen of 2,500-Ma-old conglomerate (**B**) from the South African Venterspost Fm., which contains brassy-colored pebbles of pyrite (**C**) that would have weathered away had oxygen been plentiful.

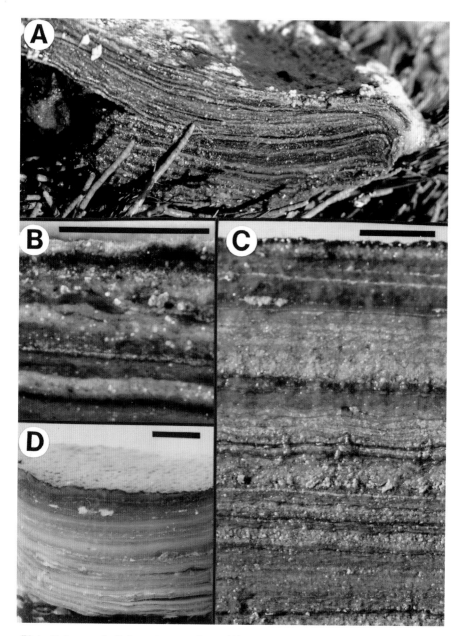

Plate 7. Layers in living stromatolites. (**A**) The upper 5-cm-thick part of a stroma-tolite from a saline lagoon in northern Baja, Mexico, sliced vertically to show its lay-ered structure. (**B**) The uppermost layers of (A) are made up of a cyanobacterial (bluish green) growth surface and photosynthetic bacterial (pink and gray-green) un-dermat, a type of layering shown also in stromatolites from (**C**) a super-salty pond in southern Baja (courtesy of D. J. Des Marais, Ames Research Center, NASA) and (**D**) a hot spring at Yellowstone National Park, Wyoming (courtesy of D. M. Ward, Montana State University). (Bars for scale represent 1 cm.)

Plate 8. Use of fossil stromatolites in China. (**A**) Stromatolitic floor tile (from the 850-Ma-old Northeast Red Fm.) in the lobby of Kwanchow's Pearl River Hotel; the five-Yuan note is the size of a dollar bill. (**B**) Playground slide near the Beijing Zoo made of slabs of 1,500-Ma-old stromatolitic limestone (Wumishan Fm.). (**C**) Great Hall of the People on Beijing's Tiananmen Square, home of China's Parliament, fronted by columns of gray stromatolitic Wumishan limestone that have ornate footings (**D**) of Northeast Red stromatolites.

8

Cyanobacteria: Earth's Oldest "Living Fossils"

Modes and Tempos in the Evolution of Life

Everyone knows the journalist's litany—who, what, when, where, why, how? Questions about the history of life are much the same: *What* evolved, *when*, and *how*?

The *what* of evolution, whether kinds of organisms or traits that show how life evolved from small to large, simple to complex, is answered by the paleobiologic record and the Universal Tree of Life. The *when* of evolution is more of a problem. Because paleobiology can set only minimum ages for life's branches, the timing of breakthrough events is hard to decipher. And the Tree of Life shows only the order in which the branches grew, not when they sprouted. Moreover, the pace of evolution does not follow simple rules. Some branches evolve fast, others slowly, and rates vary even within branches, often but not always fast early and slow later.

The *how*, the mode of evolution, is more complicated still. Its basic workings are "descent with modification . . . survival of the fittest." But being fittest depends on the setting. What wins the Darwinian struggle? Being largest, or smallest? Running fastest, or sitting still? Gobbling CHON with gusto, or getting by on little food? Spawning many offspring, or taking special care of just a few? Having a lengthy lifespan, or living only long enough to pass genes on to the next generation? Different traits win in different settings, and the winners of contests among the differently most fit are not easy to predict.

Darwin's Legacy and George Gaylord Simpson

Ever since Darwin, students of evolution have worked to discover all-embracing laws to tie together the myriad facts of life's history. These facts are at the same time the core and the minutia of the science. Unchanging and reproducible, they are the all-important nitty-gritty that must be met by any valid theory. But as a science progresses, facts soon become cumbersome to deal with, too numerous to remember. Even with the help of ever-faster computers it is left to the scientist to search out big-picture truths.

Attempts to weave known facts into the fabric of evolutionary laws took major strides in the 1940s when mathematical models of the way heredity operates in living populations were united with the rich evidence of the fossil record. The resulting union is known as the neo-Darwinian Synthesis. A leader of this movement was Columbia and Harvard professor George Gaylord Simpson (1902–1984), an expert on the history of fossil mammals who set out to tackle the when and how of evolution. Now, a half-century later, Simpson's insights are of proven lasting value.

The Rules of Evolution

In 1944 Simpson authored *Tempo and Mode in Evolution*, a slim volume rich with new ideas. His goal was to understand the familiar schoolbook history of life from seaweeds to flowering plants, trilobites to humans—the Phanerozoic world of sexually reproducing many-celled organisms that thrive because their specialized organs (flowers, leaves, teeth, limbs) fit hand-in-glove with their surroundings. While Simpson saw this kind of specialization as key to evolutionary success, he also knew its downside, for if a species is tied to a given setting it can be annihilated if the setting changes. Specialization and extinction are different sides of the same coin, so the history of Phanerozoic life is punctuated by extinctions, mostly local and hitting only a few species, but sometimes global and devastating.

Simpson codified what has come to be thought of as "standard evolution" played by the now well-known rules of the game—*speciation, specialization, extinction.* There was no reason for him to guess that life's earlier evolution would prove as dramatically different as it has.

Figure 8.1 George Gaylord Simpson, about 1980, at a garden party in Tucson, Arizona, celebrating Darwin's birthday. (Courtesy of Everett H. Lindsay, University of Arizona.)

In place of the plants and animals of the Phanerozoic, the world of the Precambrian was populated by a metabolic menagerie of non-sexual bacteria and archaeans. Rather than evolve at the pace normal for large organisms, many of these microbes changed not at all over astonishingly long spans of time. And instead of being specialized for local settings, members of the most successful of the ancient lineages—cyanobacteria—are generalists that flourish in an impressive range of surroundings. Rather than Simpson's rules, life of the Precambrian followed the path of *speciation, generalization, and exceptionally long-term survival.*

The Phanerozoic, the Age of Specialization and Extinction, differs sharply from the Precambrian Age of Generalization and Survival. The first signs of this were seen in the late 1960s, but it has taken three decades to amass enough evidence to prove the point. The telling facts come from studies of Precambrian cyanobacteria, monarchs of the primal world. And even though their fossil record still is far from fully known, the case can be made because lack of evolution (technically, "evolutionary stasis") can be nailed down by much less evidence than is needed to show evolutionary change.

Tempos of Life's History

In *Tempo and Mode,* Simpson introduced terms for three different rates of evolution shown by comparing the morphology (size, shape, structural makeup, and so forth) of fossils and their living relatives: (1) *tachytelic,* for fast-evolving species; (2) *horotelic,* the most common rate of change; and (3) *bradytelic,* for slowly evolving kinds of life.

Slow bradytelic evolvers are famous as living fossils. Good examples are horseshoe crabs, coelacanth fish, crocodiles, opossums, and primitive lampshells (linguloid brachiopods). Simpson pegged these as bradytelic because they belong to "groups that survive today and show relatively little change since the very remote time when they first appeared in the fossil record." So defined, bradytely is nearly identical to the concept of "arrested evolution" proposed in 1918 by the German-born American paleontologist Rudolf Ruedemann (1864–1956). Both ideas are based on the comparison of fossil and modern organisms that are nearly indistinguishable yet are separated by a hundred million years or more.

Simpson's three categories each span a range of rates centered around an average "lifetime" (from first appearance in the fossil record to eventual extinction), for tachytelic species about 1 Ma; horotelic, 10 Ma; and bradytelic living fossils, 100 Ma.

Coining a New Term: Hypobradytely

Cyanobacteria stand out as having evolved much more slowly even than bradytelic life. This is a major surprise that would never have

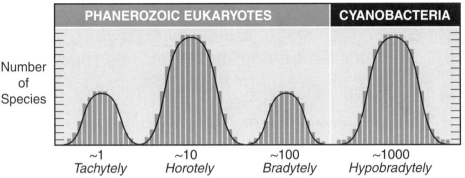

Figure 8.2 Named categories of evolutionary rates.

been guessed had the Precambrian fossil record not been uncovered. Because cyanobacteria don't fit any of Simpson's categories, some years ago I coined a new term, *hypo*bradytely, to accommodate their ultraslow change. I chose the Greek prefix "hypo" because it refers to the lowest member of a series (as in *hypo*nitrous acid, nitrous acid, nitric acid). The notion I wanted to emphasize is that these tiny ancient microbes evolved at an imperceptibly sluggish pace. Cyanobacteria—among the lowest of the low, the slowest of the slow!

This ploy of making up a fancy new term is craftier than it first appears. To a student it's a bother that there are now four labeled rates, rather than only three, to memorize for an exam. But in science, technical terms are simply shorthand notations for ideas, and as a term such as hypobradytely gains acceptance, the concept it stands for does too. As spelled out in the article that proposed the term, hypobradytely pertains "to the exceptionally low rate of evolutionary change exhibited by cyanobacterial taxa, morphospecies [named on the bases of size, shape, and cellular organization] that show little or no evident morphological change over many hundreds of millions and commonly over more than one or even two thousand million years." Simply put, the claim is that cyanobacteria maintained the status quo, changing little or not at all since they first burst onto the scene billions of years ago.

Acceptance of an Idea

It's probably not far off the mark to think of science as a system of studied skepticism. Studied, because it is a thoughtful venture carried out according to clear-cut rules. Skepticism, because what is already known works so well that a claim that doesn't fit is likely to be mistaken. Like any new idea in science, hypobradytely must be tested thoroughly.

Three tests need to be passed. First, hypobradytely must be shown to be as soundly based as Simpson's categories. (If equally grounded, it too should be accepted.) Second, it should be shown that knowledgeable workers experienced in the science agree with the notion. (Scientific concepts ought to be judged on their merits rather than popularity, but the backing of experts makes them easier to accept, especially when backing comes from the global scientific community.) Above all, of course, the idea must be validated by plenty of hard facts.

The first is easy to demonstrate. Like Simpson's categories, hypobradytely is based on *morphological* comparison of living and fossil organisms, a match made easily for cyanobacteria because all aspects of size, shape, and cellular organization used to identify living species are present also in well-preserved fossils.

The second test, popularity of the concept among experts, is equally met. Precambrian and modern cyanobacterial look-alikes have been known since 1968, when the "morphological evolutionary conservatism" of the group was first pointed out. Since then, so many fossil-modern look-alikes have turned up that it has become standard practice for the fossils to be named after their living relatives. For example, fossilized microbes that are (literally) dead ringers for living cyanobacteria of the genus *Oscillatoria* have been named *Oscillatorites* ("related to *Oscillatoria*"), *Oscillatoriopsis* ("*Oscillatoria*-like"), and *Archaeoscillatoriopsis* ("ancient *Oscillatoria*-like"). To highlight such relations, many workers have simply added the prefixes palaeo- ("old") or eo- ("dawn") to names of living genera. Nearly fifty namesakes have been proposed by workers worldwide for fossil relatives of living cyanobacteria in eight different taxonomic families.

Cyanobacterial Fossil Namesakes

CHROOCOCCACEAE
Anacystis	Palaeoanacystis
Microcystis	Palaeomicrocystis
Gloeocapsa	Eogloeocapsa
Synechococcus	Eosynechococcus
Aphanocapsa	Eoaphanocapsa
Eucapsis	Eucapsamorpha

OSCILLATORIACEAE
Lyngbya	Palaeolyngbya
Spirulina	Palaeospirulina
Microcoleus	Eomicrocoleus
Phormidium	Eophormidium
Oscillatoria	Oscillatoriopsis
Schizothrix	Schizothropsis

PLEUROCAPSACEAE
Pleurocapsa	Eopleurocapsa
Pleurocapsa	Palaeopleurocapsa

NOSTOCACEAE
Nostoc	Palaeonostoc
Anabaena	Anabaenidium

RIVULARIACEAE
Calothrix	Palaeocalothrix
Rivularia	Primorivularia

SCYTONEMATACEAE
Plectonema	Eoplectonema
Scytonema	Palaeoscytonema

ENTOPHYSALIDACEAE
Entophysalis	Eoentophysalis

HYELLACEAE
Hyella	Eohyella

Namesakes coined by scientists in Brazil, Canada, China, India, Israel, Russia, USA

Figure 8.3 Fossil cyanobacteria named after genera living today.

The third, most crucial test, the quantity and quality of supporting evidence, is also passed with flying colors, as we will now see.

The Status Quo Evolution of Cyanobacteria

How common fossils of any biologic group are may be measured by what are called taxonomic occurrences—the number of species belonging to a given group that are known to be present in formally recognized geologic units. For instance, ten species of fossil cyanobacteria in each of three named geologic formations adds up to thirty taxonomic occurrences. More than four thousand such occurrences of fossil microbes are known from Precambrian rocks. But like the fossil record of stromatolites, the microbial record is uneven. It extends to

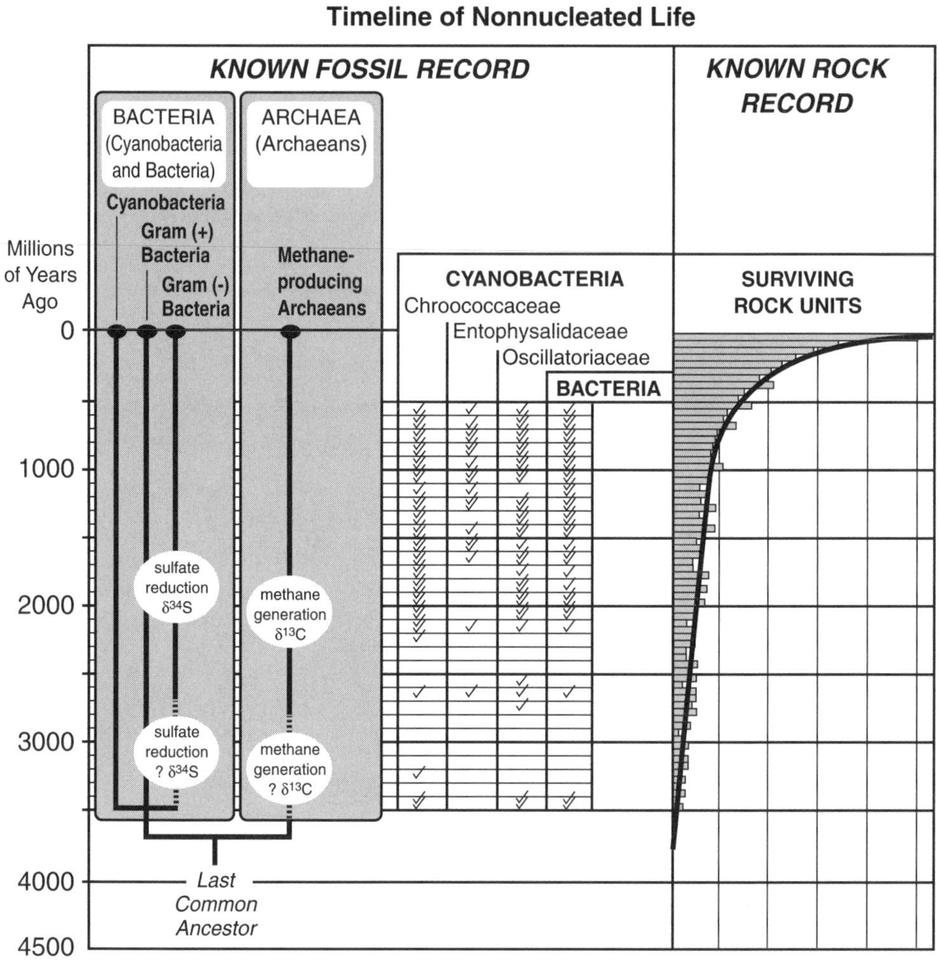

Figure 8.4 Comparison of the distributions in time of known Precambrian microbial fossils and rock units that have survived to the present. The check marks in the central columns show the spread in time of the thousands of taxonomic occurrences known from Precambrian rocks. Each check signifies that microbes of the type listed have been found in the 50-Ma-long interval indicated.

nearly 3,500 Ma ago, to the fossils of the Apex chert (chapter 3), but is meager in rock units older than about 2,200 Ma—mainly because relatively few rocks from this time have survived to the present, and most of these have experienced the fossil-destroying heat and pressure of metamorphism (but also because most workers have chosen to hunt in the younger terrains where well-preserved fossils are plentiful and it's easier to make finds).

Caveats

With more than four thousand taxonomic occurrences known, most of these of cyanobacteria, the Precambrian fossil record has now been explored sufficiently to judge whether hypobradytely makes sense or not. But the occurrences show also that the fossil evidence is spread unevenly, petering out in the most ancient terrains. This is but the first of five potential problems:

(1) *The known record of Precambrian life is spotty and incomplete, especially from earliest Earth history. Is this slim fossil record adequate to provide convincing evidence of hypobradytely?*

Even Precambrian cyanobacteria, the best-documented branch of early life, have a scanty fossil record—for ball-shaped species they amount to fewer than fifty taxonomic occurrences per 50-Ma-long interval, for filament-forming kinds fewer than two dozen. Though much more evidence would be needed to sort out rapid evolution like that of the Phanerozoic, this spotty fossil record, like occasional signposts all pointing to a single destination, is ample to show a *lack* of change, the maintenance of a status quo, over geologically long periods.

(2) *Cyanobacteria are tiny, span a limited range of shapes, display none of the complicated body plans of higher forms of life. How sure can we be that fossil and modern species are the same if there are so few features to match?*

It would be troublesome if so few traits were used to compare advanced large organisms. But for cyanobacteria this poses no problem because every feature of size, shape, and organization used to catalog living species is preserved in fossil members of the group.

(3) *Some small cyanobacteria look much like "large" noncyanobacterial microbes. Can minute fossil cyanobacteria be distinguished from bacterial look-alikes?*

Placement of some Precambrian fossils in the cyanobacteria, especially very tiny forms, is uncertain. But this caveat applies to only a few percent of the thousands of taxonomic occurrences. Cyanobacterial-bacterial mimicry poses a problem, but only a minor one.

(4) *Hypobradytely is based on the morphology of fossil and modern microbes, not biochemistry. Does similar morphology always reflect similar metabolic lifestyles?*

This question has two parts. First, do all microbes that look like cyanobacteria have their oxygen-producing photosynthetic lifestyle? Second, might cyanobacteria be afflicted by a cellular (rather than stromatolitic) version of the Volkswagen Syndrome, a lack of change in external form that masks the evolution of internal biochemical machinery?

The first part is easy to answer. The vast majority of microbes that look like cyanobacteria have the oxygen-producing and oxygen-consuming physiology of full-fledged members of the group. Aside from the possible mixup of tiny cyanobacteria and extra-large bacteria (the problem of microbial mimicry), there are fewer than a dozen living species of noncyanobacterial microbes that might cause confusion (a few non-oxygen-producing photosynthetic bacteria and chemo-autotrophic sulfate-reducers).

Though an imaginative concept, a cellular Volkswagen Syndrome seems unlikely to have happened. From early in biologic history to today, the same families, genera, and even species of cyanobacteria have inhabited the same settings, lived in the same kinds of microbial communities, and built the same thinly layered, high-rise stromatolitic condos. This remarkably stable set of relations could have been sustained only if the metabolic lifestyle of cyanobacteria remained unchanged over the eons.

(5) *Some fossil cyanobacteria are flattened and wafer thin, others three dimensional. How can these be compared and matched with living species?*

As we saw in chapters 2 and 7 (and will see again in chapter 9), the search for ancient life has advanced along parallel pathways: (1)

investigations of fossil phytoplankton preserved as compressed carbonaceous films and studied in acid-freed residues of fine-grained sedimentary rocks such as siltstones and shales (see plates 3 and 4); and (2) studies of bottom-dwelling microbes petrified in a three-dimensional, lifelike state in stromatolitic communities and studied in petrographic thin sections of shallow-water cherty rocks (shown in plates 1 and 2). Because different expertise is required for preparation and study of these different kinds of rocks, few researchers investigate both. And because compressed organic films preserved in siltstones or shales differ greatly in appearance from three-dimensional microbes, whether living today or petrified in cherts, fossils in fine-grained sediments have been given one set of names following one set of rules; those in cherts other names by other rules; and a species found in both is likely to have been named twice. The flattened fossils of fine-grained rocks are particularly vexing because happenstance can produce different-looking wrinkled, crimped, crushed, or torn variants of a single species, each likely to be knighted by a different Latinized moniker. Redundant species names have multiplied like weeds and until just recently made it all but impossible to tote up the abundance and track the diversity of early life.

(It is worth pointing out, however, that the proliferation of epithets is not just a matter of how fossils are preserved and the expertise of the finder. In the name of cost-efficiency, scientists everywhere are under pressure to increase productivity. But because intellectual accomplishments are hard to measure, bureaucrats often turn to numbers—of books authored, papers written, or new species described. Some research institutes have even set up quotas for the "number of species expected to be discovered and named yearly" by members of their staff. As scientists have obliged, named species have multiplied needlessly. Unfortunately—but not surprisingly—this mentality has seeped through to students. The coveted prize of a Ph.D. is awarded for contributions new to science, and one way to judge "newness" is by the number of new species uncovered. More than a few doctoral degrees have been garnered by overzealous students who tag even minor biologic or preservational variants with freshly minted names or rediscover, redescribe, and redub species already known.)

To sort through this chaos—to ensure that a fossil "species" is truly

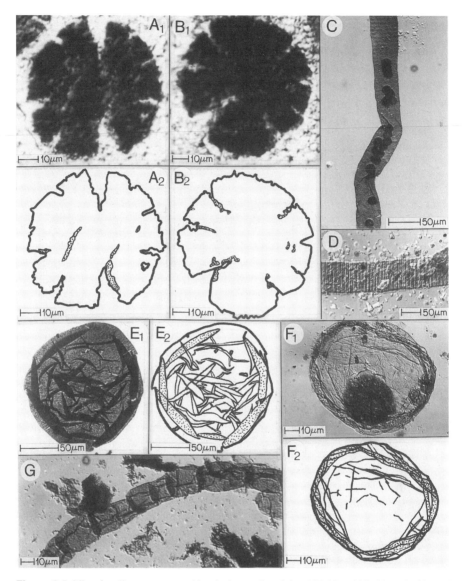

Figure 8.5 Microfossils compressed in shaley rocks of the 950-Ma-old Lakhanda (**A** and **B**) and 850-Ma-old Miroedikha Formations (**C** and **G**) of Siberia, and the 580-Ma-old Redkino (**D** and **F**) and 800-Ma-old Akberdin Formations (**E**) of Bashkiria.

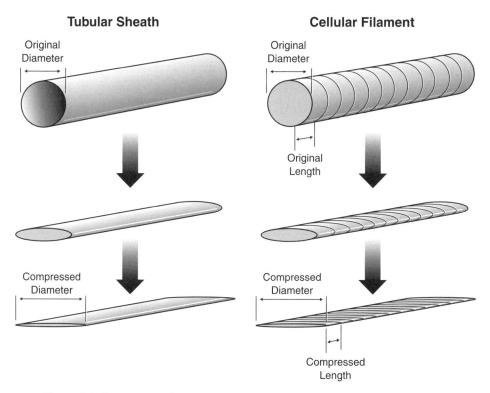

Figure 8.6 Comparison of three-dimensional and flattened fossil filaments.

a species and "new" ones are actually new—means have been devised to match forms flattened in shales with living microbes and the
lifelike fossils preserved in cherts. The important (and simple) idea is
that a fossil filament compressed between layers of mud or silt
responds like a thin-walled garden hose. As it is leveled to wafer
thinness, its width spreads evenly and this makes its original size easy
to reconstruct from the width of the squashed strand. Ball-shaped microbes flatten like punctured balloons. If their walls don't tear, they
compress to ultrathin fossil disks the same diameter as the original
sphere-shaped cells.

Now finally set straight, these relations have opened the way to
study the early fossil record of siltstones, shales, and cherts together,
not piecemeal as was the practice, and to compare the fossils in each

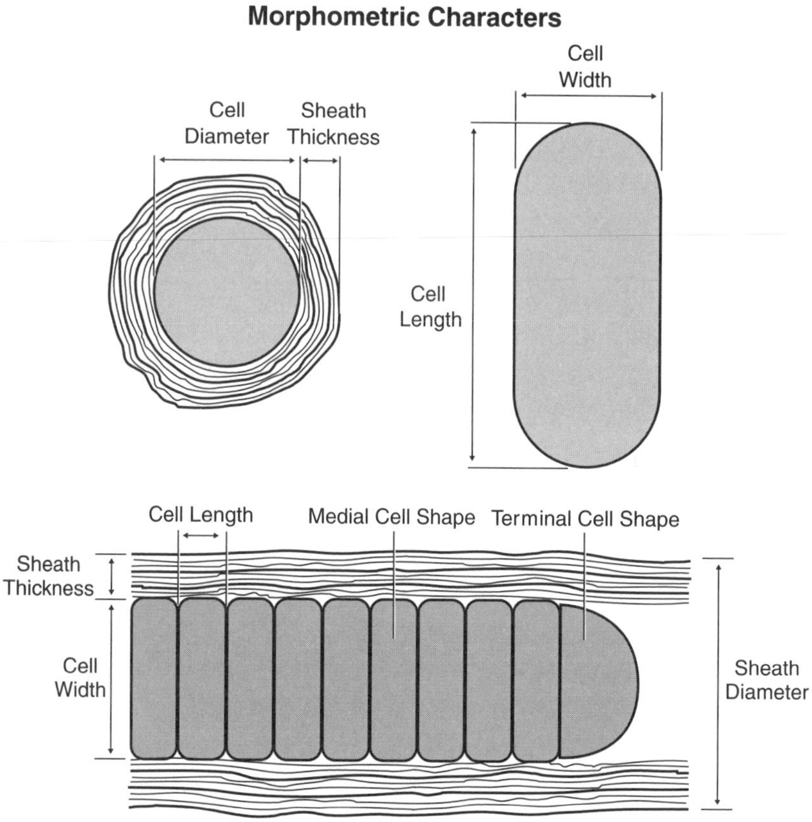

Morphometric Characters

Figure 8.7 Traits measured to compare modern and fossil microbes.

with microbes living today. The comparisons are made by measurement of morphologic traits, a quantitative technique called "morphometric analysis."

Morphometrics and "Probable Identity"

Ten traits of shape and size can be used to compare fossil and modern microbes. Living forms have telltale biochemistries as well—distinctive enzymes, lipids, photosynthetic pigments, and DNA and RNA base sequences—but in fossils these are long lost, altered by heat and pressure and leached away by groundwater and petrifying fluids. Morphology is only part of the database used to identify living microbes,

MEDIAN DIMENSIONS

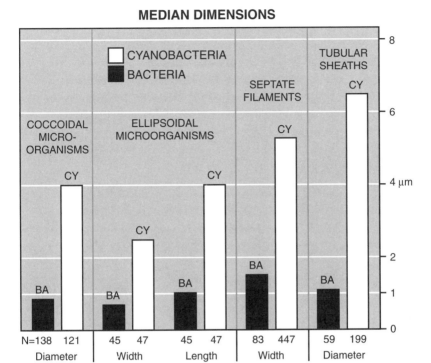

Figure 8.8 Median cell sizes of various types of living cyanobacteria and noncyanobacterial bacteria.

but it provides good grounds to distinguish bacteria from archaeans and most cyanobacteria from other members of the Bacterial domain.

Almost all fossil microbes are similar in size, shape, and cellular organization to those living today. And because these traits vary within gene-determined ranges that earmark modern groups, they can be used to identify fossil relatives. Cell size, easy to determine even in flattened fossils, is especially suitable for fossil-modern comparisons. Most cyanobacteria are made up of cells several times larger than those of similar-shaped noncyanobacterial prokaryotes, a difference used to class large-celled fossil microbes as cyanobacteria, tiny ones as noncyanobacteria. But intermediate-sized cells can pose a problem. A pattern typical of most morphologic traits is the overlap of cell size between ball-shaped cyanobacterial and noncyanobacterial

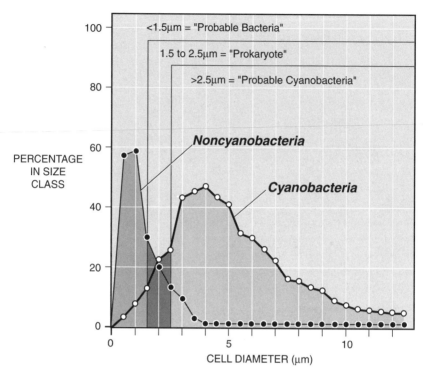

Figure 8.9 Overlapping size ranges of living spheroidal cyanobacteria and non-cyanobacterial bacteria.

prokaryotes. Cyanobacteria tend to be large, mostly 3 μm or more, and noncyanobacteria minute, 0.5 to 1 μm, but both groups include species 1.5 to 2.5 μm across, a range where they cannot be told apart by size alone. From this comparison, spheroidal fossil microbes larger than 2.5 μm are likely to be cyanobacteria, those smaller than 1.5 μm noncyanobacterial bacteria. Fossils in the 1.5- to 2.5-μm twilight zone could belong to either group and are properly referred to simply as (undifferentiated) prokaryotes.

Some morphologic traits are present only in cyanobacteria, others in noncyanobacterial bacteria, still others in archaeans. Fossil microbes having these traits are easy to classify, and though the identity of those having morphologies that overlap between the various groups is less certain, most can be placed with confidence in one or another of life's major branches.

The Prokaryotic Fossil Record

As we saw in chapter 7, the most common Precambrian fossils are ball-shaped cells and stringlike filaments. Some strings are made up of single file rows of simple uniform cells, others are hollow spaghettilike tubes. Abundant in nearshore cherty stromatolites, the balls and strings are common also in siltstones and shales, especially those laid down in coastal lagoons and drying mudflats.

The ball-shaped fossils occur singly, in pairs, or in colonies of a few, hundreds, or even thousands of cells and are often surrounded by one or more layers of wispy organic film, remnants of encasing mucilage envelopes. Their many kinds of colonies are all found among living members of the Bacterial domain—irregular masses, globe-shaped clusters, spheroidal rosettes, rectangular sheets, tiered cubes. The smallest-celled varieties are probably noncyanobacterial microbes, but most can be placed in the cyanobacterial family Chroococcaceae.

The segmented or tubular stringlike filaments are unbranched; straight, curved, twisted, or coiled; and often interlaced in feltlike mats a millimeter or so thick that make up the stacked layers of stromatolites or the thin, microbial veneers of mudflats. Narrow filaments are often long and sinuous, whereas thicker strands tend to be shorter, preserved as broken, stubby fragments. The segmented strings are made up of cells uniform in size and shape except at their ends, where they are capped by terminal cells that often have a different shape. A typical example, the cyanobacterium *Phormidium*, can be visualized in the mind's eye as a chain of tiny softdrink cans (cylindrical cells) with cone-shaped gumdrops (terminal cells) tacked on both ends, a cellular chain called a "trichome." Switch the cans to 5¢ coins and round off the ends, and the trichome form changes to that of *Oscillatoria*. Encase the chain of coins with a cylinder of plastic food wrap (the enveloping tubular mucilage sheath) and the form becomes like a sheath-enclosed trichome of *Lyngbya*, an outer tube and inner cellular chain that together are known technically as a "filament." Take away the coins, leaving only the food-wrap cylinder, and the filament becomes an empty, transparent tube like the gauzy sheaths left behind when cellular trichomes glide away to seek sunlight for photosynthesis.

Some narrow cellular or tubular fossils are remnants of noncyano-

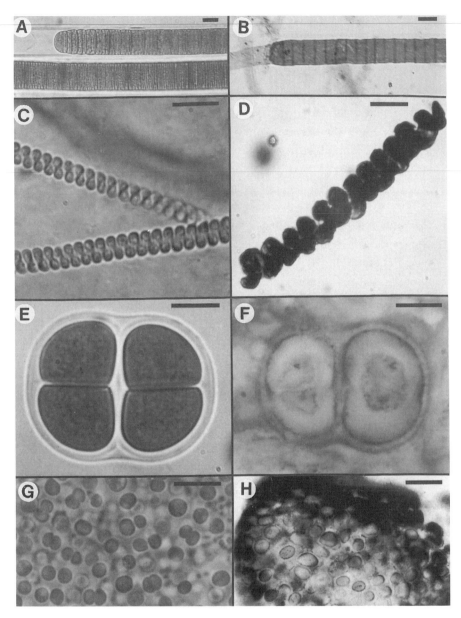

Figure 8.10 Living stromatolite-building cyanobacteria from northern Mexico (A, C, E, and G) and Precambrian look-alikes (B, from the 950-Ma-old Lakhanda Formation, and D, the 850 Ma-old Miroedikha Formation, both of Siberia; F, the 1,550-Ma-old Satka Formation of Bashkiria; and G, the 2,100-Ma-old Belcher Supergroup of Canada). (**A**) *Lyngbya*, compared with (**B**) *Palaeolyngbya*. (**C**) *Spirulina*, compared with (**D**) *Heliconema*. (**E**) *Gloeocapsa*, compared with (**F**) *Gloeodiniopsis*. (**G**) *Entophysalis*, compared with (**H**) *Eoentophysalis*. [Bars for scale represent 10 μm. (H) courtesy of H. J. Hofmann, University of Montreal, Canada.]

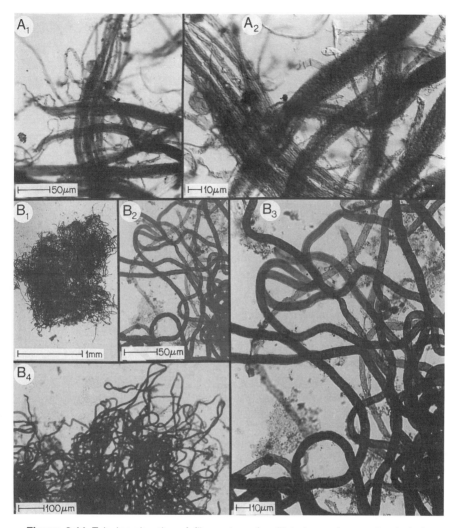

Figure 8.11 Tubular sheaths of filamentous (oscillatoriacean) cyanobacteria in acid-resistant residues of shaley rocks from the 850-Ma-old Miroedikha Formation of Siberia.

bacterial microbes, but like *Phormidium*, *Oscillatoria*, and *Lyngbya*, most belong to the cyanobacterial family Oscillatoriaceae. Often only the spaghettilike sheaths are preserved, partly because they are vacated when trichomes move to more agreeable suroundings but also because they are more resistant to decay than the cells of trichomic threads.

In addition to chroococcacean balls and oscillatoriacean strings, three other cyanobacterial families are fairly common in the early fossil record: the Entophysalidaceae, Pleurocapsaceae, and Hyellaceae.

Entophysalidaceans such as *Entophysalis* and its fossil look-alike *Eoentophysalis* are composed of jelly-bean-shaped cells that form lumpy, slime-embedded colonies on rocky substrates. Pleurocapsaceans such as *Cyanostylon* and its fossil counterpart *Polybessurus* are egg-shaped cyanobacteria that live in close-packed groups where they sit atop long slender gelatinous stalks that radiate upward from the seafloor in pincushion-like clumps. Hyellaceans, represented in modern settings by *Hyella* and the early fossil record by *Eohyella*, are endoliths, cyanobacteria that etch tiny caves in limestone pebbles, boulders, and stony pavements which they then inhabit, living *within* the outermost rock rind where sunlight penetrates.

All five cyanobacterial families—the Chroococcaceae (balls), Oscillatoriaceae (cellular and tubular strings), Entophysalidaceae (jelly beans), Pleurocapsaceae (stalked eggs), and Hyellaceae (cavity-inhabiting endoliths)—display status quo, hypobradytelic evolution.

Fossil-Modern Counterparts: Balls and Strings

Hypobradytely is especially well documented for the balls and strings, the most common kinds of cyanobacteria in the early fossil record. Comparison of more than 600 species of living cyanobacteria with a worldwide sample of Precambrian fossils, both chroococcacean (nearly 2,000 taxonomic occurrences in about 300 geologic formations) and oscillatoriacean (750 occurrences in 200 formations), shows that practically all of the fossils can be placed in present-day genera, and up to 40% cannot be told apart from particular living species. All colony forms known in the modern groups are present among the fossils, and the fossil tubular sheaths are identical in shape, size, and detailed structure to those of living counterparts. Most of the several hundred fossil-bearing rock units were laid down in coastal lagoons, mudflats, and tide-washed shallow platforms. From early in the Precambrian to today, these same settings have been inhabited by the same suite of chroococcacean and oscillatoriacean cyanobacteria.

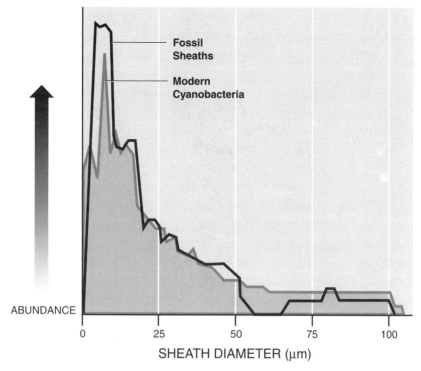

Figure 8.12 Comparison of the size of the tubular sheaths of living oscillatoria-cean cyanobacteria with those in the Precambrian fossil record.

Jelly Beans, Stalked Eggs, and Cavity-Inhabiting Endoliths

Status quo evolution is typical of other cyanobacteria as well, including distinctive jelly-bean-shaped members of the Entophysalidaceae. One of several excellent examples is the Canadian 2,100-Ma-old genus *Eoentophysalis*, which is all but identical to modern *Entophysalis*—in cell shape, colony form, the way the cells divide and grow, the stromatolitic structures they build, the environments they inhabit, the makeup of their microbial communities, even the way their cells break down when they die.

Egg-shaped pleurocapsaceans also changed little over time. Fossils known as *Eopleurocapsa* and *Paleopleurocapsa* are indistinguishable from species of the modern genus *Pleurocapsa*. And pincushion-like clumps of the stalked, egg-shaped cells of *Polybessurus*, known from 770-Ma-old stromatolites of South Australia and East Greenland,

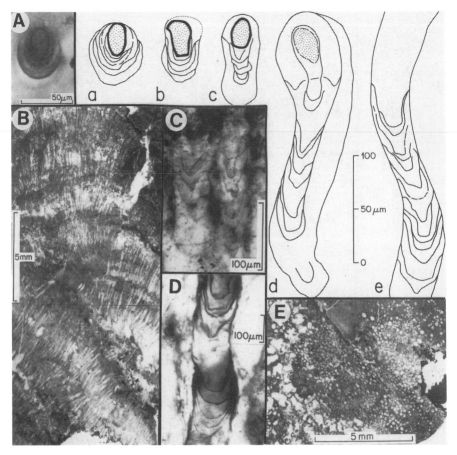

Figure 8.13 The colonial, pincushion-forming fossil cyanobacterium *Polybessurus* from the 770-Ma-old Skillogalee Dolomite of South Australia, showing in parts (a) through (e) stages in its life cycle. (**A**) Single egg-shaped cell. (**B**) Vertical and (**E**) horizontal sections through the fossil colony. (**C** and **D**) Preserved mucilagenous stalks like those reconstructed in parts (d) and (e).

have the same morphology, reproduction, and pattern of growth as the pleurocapsacean *Cyanostylon* living today in coastal waters of the Great Bahama Bank, the same environment inhabited by the fossil look-alike.

Further evidence is found in *Eohyella*, a Proterozoic hyellacean cyanobacterium known for its unusual rock-boring endolithic lifestyle and described by its discoverers as a "compelling example of the close resemblance between Proterozoic prokaryotes and their modern counterparts," a fossil "morphologically, developmentally, and behaviorally indistinguishable" from living *Hyella* of the eastern Caribbean.

The Bottom Line

Cyanobacterial hypobradytely is backed by an impressive body of scientific data, the nearly identical organismal form, cellular size and shape, growth characteristics, behavioral traits, and environmental settings shared by hundreds of examples of fossil-modern look-alikes belonging to five very different families. The bottom line is undeniable: cyanobacteria changed little or not at all since they came on the scene billions of years ago.

This is surprising, even startling. In the Phanerozoic, bradytelic living fossils are rare, curiosities that stick out because they are at odds with standard evolution. Yet during the Precambrian, the most successful of life's early branches evolved at an almost imperceptibly sluggish rate, and for diverse members of this group hypobradytely is the rule, not the exception. Why did cyanobacteria change so little over their exceedingly long history?

Evolution's Most Successful Ecologic Generalists

Reasons underlying the unexpected stasis of cyanobacterial evolution are touched on in Simpson's pioneering *Tempo and Mode*, for though he had no way to guess what the Precambrian fossil record held in store, he was much intrigued by living fossils of the Phanerozoic. To explain their unusually slow evolution, he settled on two principal factors: large-sized populations and an ability to thrive in varied environments, both shown in spades by cyanobacteria. Of the two, he thought environmental versatility was probably most important, the capability of an organism to live in a range of settings that together would make up a "continuously available environment." He reasoned that unusually slow evolution involves "not only exceptionally low rates of [change] but also survival for extraordinarily long periods of time," and noted that "more specialized [organisms] tend to become extinct before less specialized." To put a label on these observations,

Simpson invented what he called the Rule of the Survival of the Relatively Unspecialized.

Simpson's Rule was meant for Phanerozoic living fossils, mostly animals, but it fits Precambrian cyanobacteria even better, perhaps because their glacially sluggish evolution has three rather than just two main causes.

First, all cyanobacteria are nonsexual. Because Simpson's argument centers on animals, advanced organisms that use sexual means to multiply their kind, he views the absence of sexual reproduction as unimportant for slow evolutionary change. But as we will see in chapter 9, sex greatly speeds evolution by serving up a huge supply of new combinations of genes, so its lack in cyanobacteria must have slowed the process. Still, given the long history of these microbes and even moderate rates of mutation to move evolution along, absence of sex cannot be the sole explanation for their hypobradytely.

Second, cyanobacterial populations are huge, like those of most microorganisms made up of billions, even trillions of individuals. And because cyanobacteria are so tiny they are spread easily—by swirling waters, winds, tornadoes, hurricanes—so that many species are found worldwide. Though Simpson did not figure on such gigantic cosmopolitan populations, his reasoning still applies: organisms having especially large populations evolve especially slowly because evolutionary change cannot be spread at a rapid clip.

Third, cyanobacteria can live almost anywhere. This versatility, surpassing any Simpson might have imagined, is the keystone to their success.

Cyanobacterial Versatility

Cyanobacteria are remarkable. They have the longest fossil record, changed least over geologic time, were monarchs of the living world for most of its existence, and invented oxygen-releasing photosynthesis on which later life depends. Like fantastic aliens of a class B movie, they've proven impossible to wipe out, surviving on and on as life around them has gone extinct.

The versatility of cyanobacteria is shown especially by the Chroococcaceae and Oscillatoriaceae, the balls and strings so abundant in the early fossil record. They live, even flourish, in almost total dark-

Light Intensity:	Extremely Dim (1 to 5 $mEs^{-1}m^{-2}$)	Normal Light (50 to 60 $mEs^{-1}m^{-2}$)		Exceedingly Bright (>2,000 $mEs^{-1}m^{-2}$)
Conditions:	Cultures	Optimum Growth		Intertidal Zone
Chroococcaceae:	✔	✔		✔
Oscillatoriaceae:	✔	✔		✔
Salinity:	<0.001 – 0.1%	3.5%	27.5%	100 – 200%
Conditions:	Freshwater	Marine	Great Salt Lake	Salterns
Chroococcaceae:	✔	✔	✔	✔
Oscillatoriaceae:	✔	✔		✔
Acidity/Alkalinity:	Acid (pH 4)	Neutral (pH 7 to 9)		Alkaline (ph 11)
Conditions:	Hot Springs	Optimum Growth		Alkaline Lakes
Chroococcaceae:	✔	✔		✔
Oscillatoriaceae:		✔		✔
High Temperature:	70°C	74°C	111°C	112°C
Conditions:	Hot Springs	Hot Springs	Dried	Dried
Chroococcaceae:	✔	✔		
Oscillatoriaceae:	✔		✔	✔
Low Temperature:	-269°C	-196°C	-55°C	-2 to +4°C
Conditions:	Liquid Helium	Liquid Hydrogen	Frozen	Antarctic Lakes
Chroococcaceae:				
Oscillatoriaceae:	✔	✔	✔	✔
Desiccation:	88 yr	82 yr		Absence of Rainfall
Conditions:	Dried	Dried		Atacama Desert
Chroococcaceae:		✔		✔
Oscillatoriaceae:	✔			✔
Oxygen:	<0.01%	1%	20%	100%
Conditions:	Anoxic Lakes	Blooms	Ambient O_2	Cultures
Chroococcaceae:	✔	✔	✔	✔
Oscillatoriaceae:	✔	✔	✔	
Carbon Dioxide:	0.001%	0.035%	3.5%	40%
Conditions:	Culture	Ambient CO_2	Cultures	Cultures
Chroococcaceae:	✔	✔	✔	✔
Oscillatoriaceae:		✔	✔	
Radiation:	Ultraviolet	X-Rays	γ-Rays	Highly Ionizing
Conditions:	290 – 400 nm*	200 kr†	2,560 kr‡	Thermonuclear Bomb
Chroococcaceae:	✔			
Oscillatoriaceae:	✔	✔	✔	✔

* Absorbed by scytonemin pigment in sheaths
† Twice as resistant as eukaryotic microalgae
‡ Ten times as resistant as eukaryotic microalgae

Figure 8.14 Growth and survival of living cyanobacteria of the Chroococcaceae (balls) and Oscillatoriaceae (strings).

ness to extreme brightness. In pure, salty, or the most saline waters. In acid hot springs or lakes so alkaline almost nothing else survives. In scalding ponds or frigid ice fields. In the near-absence, presence, or huge overabundance of oxygen or carbon dioxide. In the driest locale on Earth, the Chilean Atacama Desert, where rainfall has never been recorded. Even in the deadly radiation of a thermonuclear blast! Many can fix nitrogen gas, plucking N_2 out of the atmosphere and building it into protein-forming amino acids. Provided with light, CO_2, a source of hydrogen (H_2O or, for some, even H_2S or H_2), and a few trace elements, they are terrific pioneers, often the first to colonize newly formed volcanic islands. Cyanobacteria are truly unusual—if we humans had their capabilities, we would have no problem colonizing our solar system and beyond!

The success of cyanobacteria comes because they are *generalists*, able to survive and grow under the most varied conditions. They have no need to evolve, for even if they are outcompeted in a local setting they easily find refuge in places their competitors cannot endure. This jack-of-all-trades survival strategy, so different from the specialization of Phanerozoic plants and animals, begs for explanation.

Why Are Cyanobateria So Tolerant?

Life survives by fitting its surroundings. But surroundings change as the global environment evolves, changes that have been especially great for branches of the Tree of Life that date from the very distant past. Cyanobacteria adapted as the environment evolved but never lost their mastery of the settings they faced before. In the process they developed enormous versatility and were themselves the root of the greatest change ever to affect the planet—the onset of an oxygen-rich atmosphere.

Early in Earth history, when cyanobacteria first spread across the globe, free oxygen was in short supply. The O_2 their photosynthesis pumped into the surroundings was quickly scavenged and sedimented in the iron oxide minerals of banded iron formations (chapter 6). Because oxygen concentrations were low, there was hardly any UV-absorbing ozone layer, and the Earth's surface was bathed in a deadly stream of ultraviolet light. Cyanobacteria faced a quandary. They needed sunlight to power photosynthesis, but if they grew in waters too shallow to shield them from the lethal radiation they would be wiped out. They first countered this threat by living deep under water, using gas-filled cellular pockets (vesicles) to control their buoyancy and having photosynthetic machinery that operated in exceedingly faint light, a strategy used today by the cosmopolitan and exceptionally abundant marine chroococcacean *Synechococcus*.

Oxygen and ozone began to accumulate, but concentrations remained low and UV a menace. To colonize shallow-water settings, cyanobacteria invented biochemical means to repair UV-caused cellular damage and came up with other protection mechanisms as well. Single-celled and colonial ball-shaped chroococcaceans ensured the cover of overlying waters by cementing themselves to the shallow seafloor with gelatinous mucilage, some infused with a special UV-absorbing biochemical, scytonemin. Bottom-dwelling oscillatoriaceans developed the ability to glide toward or away from light, depending on its intensity, and entwined themselves in feltlike stromatolitic mats that blanketed shallow basins.

As we saw in chapter 5, the gas warfare carried out by cyanobacteria gave them a telling advantage over their oxygen-sensitive competitors in the fight for photosynthetic space. They triumphed wherever photosynthesis could occur, from the open ocean to nearshore shallows; in lagoons, lakes, seas, and streams; in frigid to blistering hot locales; from exposed mudflats to deserts and the rocky land surface. The survive-then-thrive-almost-anywhere lifestyle of these remarkable living fossils enabled them to take over the globe.

Simpson's Rule of the Survival of the Relatively Unspecialized fits cyanobacteria to a tee. Suited to an amazingly wide range of habitats there was no need for them to ever change. Some experts claim that living fossils are simply champions at warding off extinction. If so, the Grand Champions, over all of geologic time, are hypobradytelic cyanobacteria!

9

Cells Like Ours Arise at Last

Life Like Us Has Cells Like Ours

It's easy to forget that our view of the living world is biased—understandably but unavoidably—by a human-centered outlook. We feel a special bond with life like ourselves, less with that which differs. Empathy for animals is especially deep seated. Their unmistakable emotions about each other, their offspring, home territory, favorite foods, seem like our own. We're part of the animal world, too, so it's not hard to understand why activists for animal rights decry the use of animals in medical research or of furry pelts for designer clothing.

Still, it is notable that our empathy is highest for the most human-like animals. Chimpanzees and monkeys resemble us in many ways and dogs and cats are our pets, so experiments on any of them are understandably unsettling. And fur is found only in mammals, our closest kin, so as apparel its use strikes especially close to home. But what about pigs, fruit flies, even nematode worms? All are animals. All are subjects of medical research. But the pig lobby is mostly silent, and few if any of us get upset about experiments on fruit flies or nematodes. The less like us an organism is, the less we are concerned about it. Could it be that some think it is "bad" to eat red meat, but not white because white-fleshed chickens and fish are not part of our clan? Or that vegetarians are thought to be purer still, since they eat only members of the plant kingdom?

I have no quarrel with the animal rightists, or the fowl-or-fish-onlies, or the vegetarians. My point has to do with evolution. The closer an organism is to us on the Tree of Life, the easier it is for us to feel kinship. And none of our close relatives look anything like the tiny microbes of the Precambrian. So different from ourselves, life of the distant past seems foreign. But as we've seen in earlier chapters, the Tree of Life actually is peppered with tiny single-celled forms of life. Of its many branches, only three—plants, fungi, and animals—include large, many-celled organisms, and all contain microscopic ones as well. Our big-organism bias stands at odds with life's long history.

Whether large or small, living or fossil, life comes in just two varieties: (1) prokaryotes, nonnucleated microbes of the Bacterial and Archaeal domains, the only life on Earth for most of the planet's history; and (2) eukaryotes, members of the Eucaryal domain earmarked by cells like ours that have chromosomes packaged in a saclike nucleus. Eukaryotes started out as small single cells, perhaps about two billion years ago, and only much later evolved into the many-celled organisms we know so well. And though the earliest kinds seem not at all like us, it nonetheless is true that the architecture of our cells, the way they grow and multiply, even the roots of human sex began with them. Tiny, ancient single-celled eukaryotes are far closer to us than we may imagine.

DNA and Development: Keys to Eukaryotic Success

Why are eukaryotes such a successful group? The beginnings of an answer can be found in the makeup of their cells, the robust compartments that enclose various kinds of small, membrane-surrounded bodies called organelles, each of which plays its own special role in the process of living. Plant cells contain the full complement: a nucleus that houses DNA-containing chromosomes; ribosomes, where DNA-stored information is used to make proteins and other chemical compounds; chloroplasts, where sunlight is captured and the cell's food is made by photosynthesis; and mitochondria, where food is broken down and energy released by aerobic respiration.

Apart from ribosomes, prokaryotes have none of these organelles. Yet all are always present in eukaryotes, with only two exceptions: (1) chloroplasts are absent from the nonphotosynthesizers such as protozoans, fungi, and animals; and (2) mitochondria are lacking also in the deepest two eukaryotic branches of the Tree of Life, the diplo-

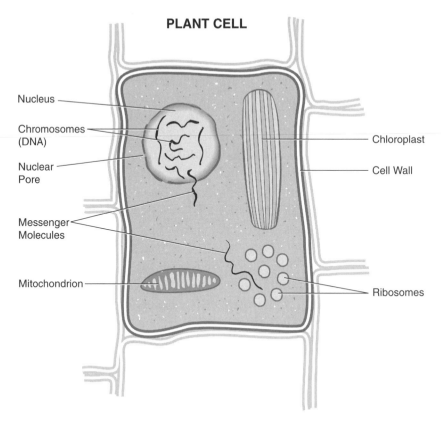

PLANT CELL

Nucleus

Chromosomes (DNA)

Nuclear Pore

Messenger Molecules

Mitochondrion

Chloroplast

Cell Wall

Ribosomes

Figure 9.1 The main organelles of a plant cell.

monads (such as *Giardia*, a disease causer fairly common in humans) and the microspordia—both branches made up entirely of parasites. The absence of chloroplasts and mitochondria in the two apparently most primitive eukaryotic groups suggests that the first single-celled eukaryotes could neither photosynthesize nor breathe oxygen.

The presence of a chromosome-containing nucleus is the defining trait of eukaryotes and the source of their name (from the Greek *eu*, true, and *karyon*, the Greek word used in biology to refer to the nucleus; eukaryotic cells are therefore "truly nucleated," whereas those of prokaryotes, from the Greek *pro*, before, and *karyon*, date from a time "before the nucleus"). In eukaryotes, only a part of the DNA, in segments of chromosomes called exons, is coded with information used to direct manufacture of enzymes and other biochemicals. In this

Genetic DNA in Various Organisms

Organism	Amount of DNA (billions of base pairs)	Percentage Having Coded Information	Information-Containing DNA (billions of base pairs)
BACTERIUM	0.004	100%	0.004
YEAST	0.009	70%	0.006
NEMATODE	0.09	25%	0.02
FRUIT FLY	0.18	33%	0.06
PLANT	0.2	31%	0.06
NEWT	19.0	3%	0.57
HUMAN	3.5	18%	0.63

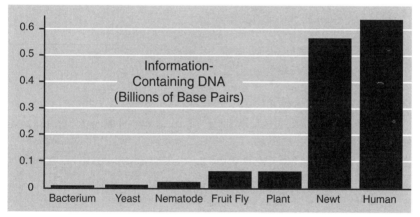

Figure 9.2 Genetic DNA in the cells of various kinds of organisms (the gut bacterium *Escherichia*; yeast, the minute fungus *Saccharomyces*; the nematode worm *Caenorhabditis*; the fruit fly *Drosophila*; the "belly plant" *Arabidopsis*; a newt, the salamander *Triturus*; and a human cell).

they differ from prokaryotes, where all of the DNA ordinarily contains instructions for making chemical compounds. But because eukaryotic nuclei house much more DNA than cells of prokaryotes, eukaryotes have a much larger store of genetic information. This helps explain their greater complexity, since how complicated an organism is depends partly on the amount of information-bearing DNA its cells possess. For instance, the cells of fairly simple organisms such as bacteria and yeast, a kind of fungus, contain only a tiny fraction of the information-containing DNA in a human cell. Cells of nematode worms, fruit flies, and the Mouse-ear Cress (*Arabidopsis*, called a "belly plant" because it is so small that one has to lie on one's belly to get a good look at it) have only a bit more DNA, and these organisms are not especially complex, either.

But complexity is not governed solely by the amount of DNA, whether it is information containing or not. Each cell in a newt (the salamander *Triturus*) contains more than five times as much DNA as a human cell and nearly as much information. And the cells of some other eukaryotes—including plants such as the lily, *Fritallaria*, and even early-evolved backboned animals such as the lungfish, *Protopterus*—have as much as forty times more DNA than a human cell, in the lungfish housing nearly twice as much of the kind that carries information. Eukaryotes are set apart from all other life by having DNA housed in nuclei, but how complex they are does not come simply from having the DNA packaged, nor from its total amount, nor even from its having a large content of information. Instead, it comes from the way the information is put to use as an organism grows and matures, a complicated and intricately choreographed sequence of steps we call "development."

Included under this umbrella are such wonders as the transformation of a fertilized egg to a wiggling tadpole to a full-grown frog, or a wormlike larva to an iridescent butterfly. Even in humans, baby teeth come in, fall out, and adult ones take their place. A full head of hair sprouts, then from some of us it mysteriously falls away. And when hormones kick in at puberty, the changes are all but miraculous. These, like all events in development, are preprogrammed. Their turn-on and turn-off times are governed by the DNA of the nucleus, but in ways we don't fully understand.

The breakthrough advances of complex programmed development happened late in evolution, as large eukaryotes arose near the close of the Precambrian. But the stage was set much earlier, at least two billion years ago, when small nucleated cells invented a process called mitosis, a new way to multiply their kind.

How Old Are the Eukaryotes?

As we saw in chapter 4, the rRNA evolutionary tree pegs parasitic diplomonads and microsporidia as the most primitive eukaryotes living today. Judging from this, the earliest eukaryotes can be guessed to have been small unicells that lacked chloroplasts and mitochondria, so were heterotrophs rather than autotrophs and anaerobes rather than aerobes. This fits the theory of evolution by endosymbiosis (chapter 5) according to which photosynthesizing aerobic eukaryotes came later, fully equipped with engulfed chloroplasts and mitochondria gained via prepackaged evolution. But the earliest eukaryotes must have been free living, unlike either of the modern parasitic groups, and it is still not certain that among living eukaryotes diplomonads and microsporidia are truly the most primitive (rather, for instance, than being aberrant members of the much later-evolved fungi, as some experts argue).

It is also uncertain exactly when eukaryotes arose. Paleobiology can place only minimum ages on branchings of the Tree of Life, so although fossils of fully equipped eukaryotes date from about 1,800 Ma ago, the group must have existed earlier. The earliest kinds probably looked almost the same as their prokaryotic ancestors, differing mainly by having a nucleus. But because nuclei almost never leave a trace in fossils, there is no good way to identify the first true eukaryotes. All we can say for sure is that eukaryotes evolved from archaeal ancestors, and since (as we saw in chapter 6) these were present 2,800 Ma ago, or earlier, eukaryotes might have existed then too.

How Do We Know Fossil "Eukaryotes" Are Eukaryotes?

As we saw in chapter 8, the cells of ball-shaped prokaryotes are tiny, almost all smaller than 5 μm. Few are as large as 10 μm, and there are only two "large" species, both less than 60 μm. Cells of eukaryotes ordinarily are much bigger—tens, hundreds, even thousands of microns across. Other than cell size, telltale traits of eukaryotes are rare in ancient unicellular fossils, so size has come to be the chief guide to identify their remnants: cells in the 10 to 60 μm range are pegged as "possible eukaryotes," those larger as "assured eukaryotes." The 60 μm boundary seems a safe limit. No prokaryotic unicells are bigger and practically none come close.

Hints of large-celled eukaryotes are present in rocks as old as 2,100 Ma in the form of spirally coiled, millimeter-wide ribbons called *Gry-*

Figure 9.3 The enigmatic ribbonlike fossil *Grypania*, shown here from the 1,300-Ma-old Rohtas Formation of central India. Other, less well preserved specimens in rocks as old as 2,100 Ma are regarded by some scientists as early (probably algal) eukaryotes. (Bar for scale represents 1 cm. Courtesy of B. N. Runnegar, University of California, Los Angeles.)

pania preserved as imprints on rock surfaces that show the outlines of the fossil bodies but not the size of their cells. Some workers think these are remnants of single-celled tubes, more or less like the hollow stalks of a eukaryotic seaweed (*Acetabularia*) living today. But others point out their resemblance to spaghettilike strands made by colonies of the prokaryotic cyanobacterium *Nostoc*. They may be fossils of eukaryotes. Or prokaryotes. Or even of extinct forms not closely related to any living group. Unless clear-cut cells are found to decide the issue, their identity will remain in limbo.

We do know that full-fledged eukaryotes were present by 1,800 Ma ago. Probably the best examples, simple baloonlike unicells, come from recent studies of rocks taken from a steep-sided valley along a winding river near the farm town of Jixian east of Beijing, China. But similar fossils of the same age were found in Russia and the Ukraine years ago by B. V. Timofeev, one of the pioneers in the field (whom we met in chapter 2). And the Jixian specimens, though smaller and much older, are not too different from C. D. Walcott's *Chuaria*, the large single-celled eukaryote unearthed in Arizona's Grand Canyon in 1899 and the first Precambrian cellular fossil ever discovered. The existence of eukaryotes at this time is shown also by the presence in rocks of this age of organic compounds called "steranes," the altered remnants of substances (sterols) that are abundant in the membranes of eukaryotes but not present in prokaryotes.

Because the Jixian and similar-aged unicells are large, some more than 200 μm wide, there is no doubt they are fossil eukaryotes. And since the preserved cells are scattered across an ancient seafloor, we know they are algal phytoplankton, eukaryotes fully equipped with chloroplasts and mitochondria. But their exact relations to living algae are uncertain. Though they may belong to the red, green, brown, or golden-brown algae of modern seas, they might be members of an early-evolved group that did not survive to today. To acknowledge this uncertainty they have been formally named acritarchs (from the Greek, *akritos*, "confused, uncertain"). Some (acanthomorph acritarchs) are spiny, but the earliest—and in the Precambrian the most common—are smooth or pockmarked simple spheroids known as sphaeromorph acritarchs (see plate 4).

Eukaryotes Perfect the Art of Cloning

Like certain single-celled microalgae living today (some species of *Chlorella*, *Chlorococcum*, and their relatives), early acritarchs multiplied by mitosis, or body cell division. This is the simplest way for a eukaryotic cell to reproduce—a parent cell merely clones itself into two exact copies. Most prokaryotes multiply by a similar process— cell fission—in which the single chromosome in a parent cell is duplicated and the copies are passed to daughter cells formed as the parent splits in half. But cloning by mitosis is more complicated. The cells of eukaryotes ordinarily have many chromosomes, cordoned off

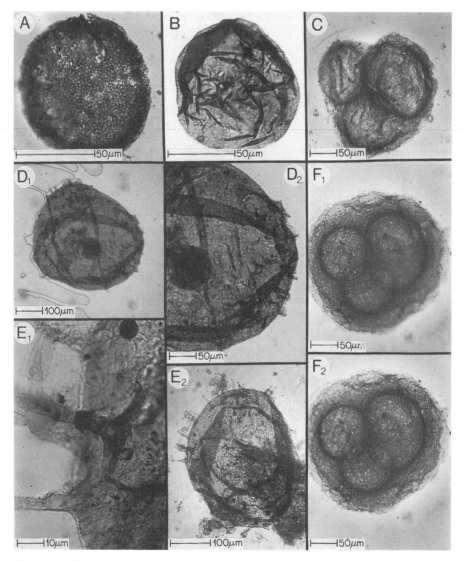

Figure 9.4 Flattened, originally spheroidal microfossils in acid-resistant residues of siltstone from the 950-Ma-old Lakhanda Formation of Siberia, in (D), (E), and (F) shown in two views of single specimens. (**A**) *Trachysphaeridium*; (**B**) *Kildinella*; (**C** and **F**) *Tetrasphaera*; (**D**) *Nucellohystrichosphaera*; (**E**) *Trachyhystricosphaera*.

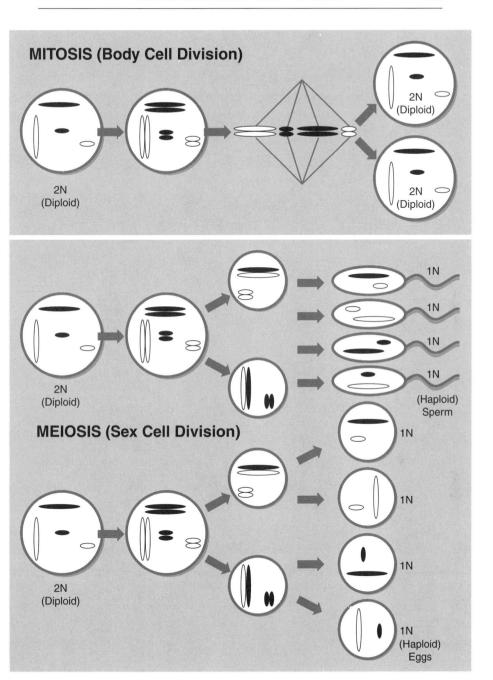

MITOSIS (Body Cell Division)

2N (Diploid)

2N (Diploid)

2N (Diploid)

2N (Diploid)

1N

1N

1N

1N

(Haploid) Sperm

MEIOSIS (Sex Cell Division)

1N

1N

1N

1N (Haploid) Eggs

2N (Diploid)

Figure 9.5 Mitosis, body cell division; and meiosis, sex cell division.

in a nuclear sac that must be broken down before the copied chromo-somes can be passed on to offspring. The chromosomes are dupli-cated, freed from the nucleus and aligned near the center of the cell, then pulled apart by protein-rich fibers (the "mitotic spindle") into two new cells, each a clone of the parent. Mitosis is a highly orga-nized yet simple way to make new cells that are faithful copies of the old.

Evolution's Goal Is to Avoid Evolving

In the mind's eye we imagine life's history as a gradually unfolding evolutionary parade from small to large, simple to complex—evolu-tion and change seem to go hand in hand. Actually, however, the aim of life is never to evolve, never to change at all! Life's slogan might well be "if it's not broken, don't fix it," and when systems do break down, for instance by mutation, life's response is immediately to re-pair the altered part back to what it was before. We see change only because biologic history spans such an exceedingly long time that rare, unfixed mutations add up.

Rather than change, life's goal is to maintain the status quo, and eukaryotic mitosis—like prokaryotic fission a near perfect sit-tight strategy—fits right in since mitotic clones are exact copies of the cell that gave them birth. Primitive microalgal acritarchs, like hypo-bradytelic cyanobacteria, evolved at an almost imperceptibly slow pace. Though mitotic phytoplankton were an important stage-setter for later evolutionary advance, the earliest, most primitive kinds changed little over many hundreds of millions of years.

Sex: A New Lifestyle Brings Major Change

Practically every eukaryotic cell can divide by mitosis, which not only makes new cells as an organism grows but replaces those dam-aged or aged—sun-burned skin, for instance, or cells sloughed off a dandruffy scalp. But using mitosis to reproduce entire higher organ-isms, not just their body cells, would lead to a monstrous outcome. If both the egg and sperm supplied by the parents contained full sets of chromosomes, the fertilized egg (zygote) would have double the number of them, the next generation four times as many, then eight, sixteen, thirty-two, and so on. Of course, this does not happen. Instead, egg and sperm are made in humans and other animals by a special type of cell division called meiosis that halves the number of chromosomes. The original number is put together again at fertilization to give the zygote a complete set, half from each parent. In animals, meiosis makes sex cells; in plants, spores that divide later by mitosis to produce egg and sperm.

Meiosis starts out much like mitotic cell division, its evolutionary ancestor. In the first step, the chromosomes double like they do in the first step of mitosis. But instead of being shunted to *two* new cells, as in mitosis, the paired chromosomes split once and then a second time and are distributed to *four* cells, each of which has only half the number of chromosomes of the starting cell.

The mitotically dividing body cells of advanced eukaryotes such as plants, fungi, and animals contain two copies of each chromosome, a complement known as the "diploid number" and abbreviated "2N" (where the N stands for number). In humans, each body cell contains 23 pairs of chromosomes so the diploid, 2N, number is 46. Cells formed by meiosis contain the "haploid" or "1N" number of chromosomes, half that of body cells, in humans 23. The two kinds of cell division, mitosis and meiosis, alternate during a eukaryote's life cycle. For example, in animals the mitotically dividing cells of an adult are diploid, each housing two copies of the chromosomes. Certain of these cells undergo changes that enable them to divide by meiosis to form egg and sperm, haploid sex cells called "gametes." When fertilization happens, the gametes fuse to form a diploid zygote. And the zygote then grows to an adult by mitosis and the cycle repeats.

Meiosis and the union of gametes (syngamy) during fertilization are the key processes that determine the sex of offspring. Gender is governed by genes, in humans packaged in chromosomes dubbed "X" and "Y." Like all body cells in females, those in special egg-making tissues have two copies of X. Since there are no Y chromosomes in females, the single set of chromosomes in each meiotically made egg can contain only an X. But because the sperm-making cells of males have both X and Y chromosomes, half the sperm carry an X, the

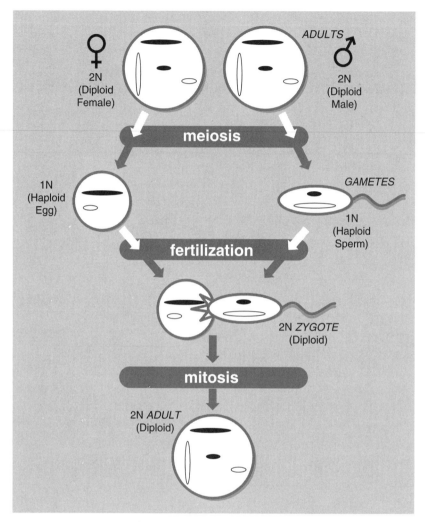

Figure 9.6 Life cycle of a sexually reproducing eukaryote. ♀, the symbol for female, represents the mirror of Venus; ♂, the symbol for male, represents the shield and sword of Mars.

other half a Y. When an X-carrying sperm fertilizes an egg, the zygote and ultimately the adult is an XX female. When an egg is fertilized by a Y-carrying sperm, the offspring is an XY male. This and systems like it explain why humans and most other eukaryotes have a 50:50 mix of the sexes.

Why Does Sex Matter?

Among all inventions evolution ever devised, only two stand out as surpassingly important: (1) oxygenic (cyanobacterial) photosynthesis, key to development of oxygen-consuming respiration and the aerobic-anaerobic workings of the modern living world; and (2) eukaryotic sex, the main source of genetic variation in higher organisms and the root of their remarkable diversity and rapid evolution.

The pre-sex living world was more or less static, evolution incredibly torpid. From time to time, new well-equipped mutants emerged among the nonsexual microbes and mitotic microalgae. But this happened only rarely because most mutations are harmful, especially useful ones are quite scarce, and reproduction by cloning maintains the status quo. Everything changed when cloning was replaced by meiotic sex, a breakthrough innovation that from the very start added a huge supply of grist to the evolutionary mill.

During meiosis, chromosomes often exchange parts to form new combinations of genes, so the suites of chromosomes parceled out vary among the gametes. Because exactly the same genes are never present in any two eggs or any two sperms, every organism born from sexual reproduction contains a genetic mix that never existed before. This is true of you, me, and everyone else! Even children of the same family always have differing combinations of genes (except for identical twins developed from a single fertilized egg). Of course, all of us have many genes in common with our sisters, brothers, and parents (and for this reason share with them what we call family resemblance), but we are not Xerox copies, not clones like Dolly the sheep.

Because the mix of genes in each individual is unique, a population of sexual organisms includes an enormous number of different gene combinations. This contrasts sharply with the monotonously uniform genetic makeup of nonsexual species and helps explain why the advent of sex was such a breakthrough. Simple examples illustrate the point. Ten mutations in a nonsexual population can produce 11 genetic combinations (technically, "genotypes"), the original and those of the 10 new mutants. But in a population of sexual organisms (assumed for simplicity to be genetically uniform except at mutant-normal gene pairs), the same 10 mutations could be shuffled to pro-

WHY NOT THREE SEXES?

Number of Mutations	Number of Potential Genetic Combinations		
	Nonsexual	Two Sexes	Three Sexes
1	2	3	4
10	11	$3^{10} = 60,000$	$4^{10} = 1,000,000$
100	101	$3^{100} = 5 \times 10^{47}$	$4^{100} = 1 \times 10^{60}$

For comparison—

One trillion stars in Milky Way galaxy = 10^{12} = 1,000,000,000,000.

Elementary particles (electrons and protons) in physical Universe = 10^{80} = 100,000,000,000,000,000,000,000,000,000,000,000,000, 000,000,000,000,000,000,000,000,000,000,000,000,000,000.

Figure 9.7 The number of genetic combinations that can be produced by a two-sex system is so enormous that additional combinations from a hypothetical three-sex system would be superfluous.

duce 3^{10} genotypes (that is, $3 \times 3 \times 3 \times 3 \times 3 \times 3 \times 3 \times 3 \times 3 \times 3$ = about 60,000). The more mutations, the greater the contrast: 100 mutations could give 101 combinations nonsexually, but by sexual reproduction 3^{100} = 5×10^{47} (five followed by forty-seven zeros).

The number of potential genotypes in the last example is astronomical, more than the stars in our galaxy (10^{12}), sizable compared even to all elementary particles in the entire Universe (10^{80}). In principle, a three-sex system could produce even more variation, but there are so many gene combinations available from two sexes that any more would be needlessly superfluous—as the Chinese say, "like painting legs on a snake."

Sex, Diversity, Specialization, and Death

By forming countless mixtures of genes, sex rapidly increased the genotypic and consequently the phenotypic (physical and biochemical) variety within species. And because numerous mixtures led to development of new species, sex also speeded the genesis of brand new kinds of organisms. But sex brought other changes as well. Most important was the division of the life cycle into two separate parts, one devoted to an organism's mitotic growth in size, the other its meiotic-based reproduction. Over time, each part became more and more adept at its particular task, and by doing so gave rise to an ever-increasing number of different kinds of eukaryotes—some specialized for growth, others for reproduction, still others proficient at both, and all suited to plumb their own special regions (niches) of the environment. As we will see in chapter 10, the prominence of sex in the Phanerozoic and its absence throughout most of the Precambrian separate decisively these two great epochs in the history of life.

The onset of sex brought yet one more momentous change—genetic death. Individual nonsexual organisms do of course die, by being eaten, poisoned, squashed, killed by disease. But their genetic makeup lives on in their cloned offspring, altered only by chance mutations. This doesn't happen in sexual organisms where the unique mix of genes in each individual departs the scene when the organism succumbs. A portion of the mix is passed to the next generation, but the same genetic makeup never exists again.

(To digress for a moment, it should be noted that this truth—the uniqueness of the genetic makeups of sexual eukaryotes—underlies the widespread concern among scientists about the human-caused extinction of species on our planet. Each species has a special mix of genes, just as each individual does—unique combinations that evolved over millions, sometimes billions of years. As species go extinct, we lose, since their genes and the biochemicals they cause to be made will no longer be present for us to learn from and use. The biochemical digitalis, for instance, a potent heart stimulant extracted from the leaves of the common foxglove plant, would not be here for our use if the foxglove went extinct. Genetic engineering holds a part of the solution. Soon we will be able to engineer almost any gene combination we can think up, so maybe we can re-create those combinations invented by organisms we have caused to go extinct. But nature's experiments have lasted eons, so we would probably be better off protecting, rather than exterminating, the gene combinations already here.)

The Wax and Wane of Precambrian Acritarchs

Balloonlike sphaeromorph acritarchs have been known since the turn of the last century when C. D. Walcott found *Chuaria* in rocks deep within the Grand Canyon. Their many species are divided into two groups, *meso*sphaeromorphs of 60 to 200 μm size, and *mega*sphaeromorphs of 200 μm or larger which include those, like *Chuaria*, that are many thousands of microns across. Both groups are known from the rocks near Jixian, China, where the oldest good specimens are found, and both are much more common in offshore shales than in shallow-water stromatolites.

Mesosphaeromorphs, the smaller variety, are found in rocks as old as about 1,800 Ma and are fairly abundant throughout the remainder of the Precambrian as well as the early Phanerozoic. The larger-celled megasphaeromorphs date from the same time but are common only in strata of the Precambrian. They evolved slowly at first, then waxed and waned as the Precambrian drew to a close.

In rocks older than about 1,100 Ma, megasphaeromorphs are rare and all are smaller than 1.5 mm (1,500 μm). The group then burst forth, and new kinds appeared rapidly, by about 900 Ma ago including forms larger than a centimeter (10,000 μm). This explosive rise was followed by a slow but steady decline, and by 300 Ma later all of the gigantic megasphaeromorphs and most of the others had died away, leaving only a few small-celled species to survive into the Phanerozoic. Spiny acritarchs suffered the same fate over the same period. Species nearly a millimeter wide are present in rocks about 950 Ma old; fewer and progressively smaller kinds occur in strata laid down from then to 600 Ma ago; and the largest in 550-Ma-old sediments of the earliest Phanerozoic are only 75 μm across.

Why the Wax?

The fossil record of Proterozoic acritarchs raises three main questions: (1) Why did eukaryotic microalgae evolve so slowly between about 1,800 and 1,100 Ma ago? (2) Why did their evolution then speed so greatly? And (3) why, after reaching their zenith at about 900 Ma ago, did they gradually go extinct, only a few small-celled kinds lingering into the Phanerozoic?

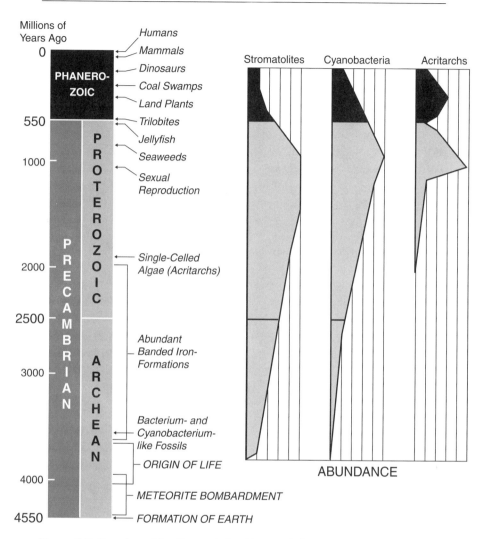

Figure 9.8 Overview of the Precambrian history of life showing (*at right*) the Proterozoic wax and wane of eukaryotic acritarchs.

The first is answered fairly easily. The advent of fully equipped aerobic eukaryotes 1,800 Ma ago or somewhat earlier was made possible by the onset of a stable oxygen-rich environment. Like certain single-celled microalgae today, these were nonsexual forms that multiplied by mitosis, or self-cloning. Though this is a simple and efficient way to reproduce, the cells formed were copies of the parent

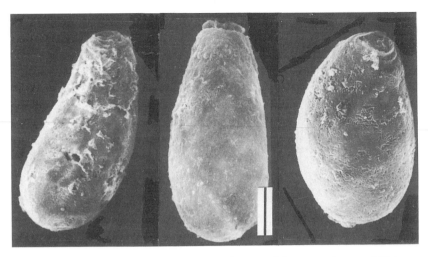

Figure 9.9 Scanning electron microscope pictures of three specimens of *Melano-cyrillium*, the thick-walled reproductive resting cells (cysts) of testate amoeboid protozoans, in acid-resistant residues of chert from the 850-Ma-old Kwagunt Formation of Arizona. (Bar for scale represents 25 μm.)

that, except for chance mutations, contained no new genetic information to spur evolutionary change. Like nonsexual hypobradytelic cyanobacteria, mitotically cloned eukaryotes evolved exceedingly slowly over hundreds of millions of years.

The burst in evolution beginning about 1,100 Ma ago seems best explained by the advent of meiosis—sex cell division—and the spread of eukaryotic sexuality, "the urge to merge." An upgrade of more primitive mitotic cell division, meiosis was in place at least as early as about 950 Ma ago, the approximate age of the oldest sexual protozoans (*Melanocyrillium*, robust reproductive cysts of walled amoebae), acritarchs having telltale wall openings (technically, "pylomes") for release of reproductive cells, and sexually reproducing small seaweeds (red algae). Compared to nonsexual eukaryotes, sexual ones would have evolved much faster, soon to include many new types. Since the fossil record of 1,100 to 900 Ma ago shows just this type of burst, the advent of sex seems a sensible and likely correct explanation for this explosive phase of early evolution, the "wax" of the waxing and waning of Precambrian eukaryotic history.

Why the Wane?

Though still an unsolved puzzle, the wane of unicellular eukaryotes, the decline and collapse of the world's microalgal flora during the 300-Ma-long period beginning about 900 Ma ago, is coming into focus. A variety of factors seem to have played a role, most prominently environmental change. This was an extraordinary time in Earth history, marked by large fluctuations in climate, at least four great ice ages, and biologically important shifts in atmospheric carbon dioxide and oxygen.

The four great episodes of global freeze are probably the result of a "reversed greenhouse effect," a change of climate opposite from that happening today. Power to run modern industry comes mainly from the burning of fossil fuels (coal, oil, natural gas), which releases energy as the carbon in the fuel is oxidized, or combined with oxygen (O_2 + fossil carbon $\rightarrow H_2O + CO_2$ + energy). This process is chemically much the same as the aerobic respiration that powers our cells and like it consumes oxygen (O_2) from the atmosphere as it gives off carbon dioxide (CO_2). The amount of free oxygen in the atmosphere is so large (more than a billion trillion grams, about 21% of the atmosphere's mass) that the drop caused by burning fossil fuels is nearly imperceptible. But because the amount of CO_2 is very much less—only 0.03% of the atmosphere, roughly 300 parts per million (ppm)—that added from fossil fuels has weighty consequences.

Carbon dioxide is called a greenhouse gas because, like the windows of a greenhouse, it holds in heat, storing it in the chemical bonds that knit its atoms together. Gases trapped in the Greenland ice sheet show that the concentration of CO_2 in the atmosphere was about 280 ppm before the beginnings of the Industrial Revolution in the late 1700s. In the two centuries since, the amount has risen to more than 360 ppm, an increase of nearly 30%. The burning of fossil fuels has pumped ever greater amounts of CO_2 into the environment, more and more heat has been stored, and the world has entered a period of "global warming."

The great ice ages of the late Precambrian are likely a result of the opposite of this effect, "global cooling," characterized by lowered temperatures brought on by *removal* of CO_2 from the atmosphere-ocean system due to its combination with calcium in seawater to form the calcium carbonate minerals of sedimented limestone rocks. Over time, the amount of heat-storing CO_2 in the atmosphere came to be less and less and the climate became so cold that massive glaciers and icebergs formed even in the tropics. Of the four episodes of continental glaciation 900 to 600 Ma ago, the two youngest were the most severe—in fact, the most severe ever: the Sturtian about 720 Ma ago, and the Varangian, a worldwide deep freeze that lasted from about 610 to 590 Ma ago, an episode that went on for *millions* of years (in constrast to the most recent, the Pleistocene ice age, which lasted only a few tens of *thousands* of years).

While CO_2 decreased, the oxygen content of the atmosphere seems to have increased, a result of cyanobacterial and algal photosynthesis followed by rapid burial of the organic matter produced. As we saw in chapter 5, in cyanobacterial and algal photosynthesis, oxygen is given off as hydrogen is split from water and combined with carbon dioxide to make glucose sugar (CO_2 + H_2O → "CH_2O" + O_2). Carbon has an atomic weight of 12 and oxygen 16, so for every 12 grams of carbon built into glucose, "CH_2O," 32 grams (2×16) of O_2 are pumped into the environment. Aerobic respiration (O_2 + "CH_2O" → H_2O + CO_2) is the reverse of this process, so 32 grams of oxygen are used up for every 12 grams of carbon that are eaten and burned by aerobic recyclers.

Aerobic respiration—breathing—is an oxygen sink that, like the iron of banded iron-formations (chapter 6), sponges up oxygen from its surroundings. If all the carbon in dead carcasses is converted each year to carbon dioxide by aerobic recyclers, there can be no change in atmospheric oxygen—the same amount of oxygen given off when the organic matter was made is used up as the carbon is recycled. And since in the world today practically all the carbon in animals and other heterotrophs comes ultimately from oxygen-producing plants and plantlike microbes (cyanobacteria), it doesn't matter whether the recycled organic matter is animal or vegetable. Only a tiny amount of the carbon in dead carcasses escapes biological recycling to be buried in sediments. Though one might think even this minute fraction would lead to a gradual buildup of atmospheric oxygen (adding 32 grams of oxygen for every 12 grams of carbon buried), it actually has no lasting effect because an equivalent amount of oxygen is consumed by the oxidation of fossil organic matter that is exposed to the atmosphere by weathering. The oxygen content of the present-day atmosphere is held at a more or less constant steady state.

A similar set of carbon-oxygen relations seems to have held atmospheric oxygen at a low, more or less steady level throughout much of the Proterozoic. As the Precambrian drew to a close, however, the evolution of a new way to bury organic carbon rapidly changed the relations, and the level of oxygen increased. Toward the end of the Precambrian, perhaps 800 or 700 Ma ago, colonial protozoans gave rise to many-celled animals. No one knows exactly when because the earliest ones were wormlike creatures so tiny that they left no trace. But some of these lived by eating muck, digesting bits and pieces of cyanobacteria and microalgae in mud and grit that passed through their bodies as they burrowed through sediments. To do this efficiently requires a gut housed within a tubular space (technically, a "coelom"), and as animals having coeloms became increasingly common so did their fecal refuse, the debris passed through their alimentary canals. Because the refuse and the undigested organic matter it contained were pelletized, bound in a mucous coating, this was a way to bury carbon rapidly. The more coelomate animals, the more fecal pellets; the more pellets, the more organic matter buried; the more buried, the more oxygen added to the atmosphere.

As CO_2 decreased, O_2 increased, and global temperatures dropped between 900 and 600 Ma ago, the oceanic environment came to be less and less agreeable to large-celled microalgae. The enzyme that drives algal photosynthesis, RUBISCO, is called a carboxylase/oxygenase because CO_2 and O_2 compete to interact with it. Laboratory experiments show that if microalgae are given decreased CO_2 and increased O_2, photosynthesis rapidly diminishes, eventually halts, and all growth stops. Bigger-celled microalgae are especially vulnerable because they need to make larger amounts of glucose to stay alive. This suggests an explanation for the late Precambrian demise of large-celled microalgae—CO_2 dropped, O_2 rose, and the key enzyme of

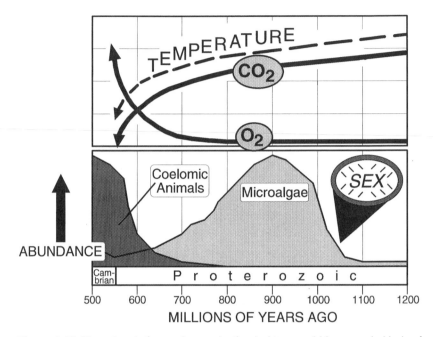

Figure 9.10 The advent of sexual reproduction led to a rapid increase in kinds of eukaryotic microalgae (acritarchs) between about 1100 and 900 Ma ago. From then to the close of the Precambrian, microalgal diversity steadily declined as atmospheric CO_2 decreased and the climate became colder. Many-celled animals arose during this period, possibly 800 to 700 Ma ago, and the coelomic kinds that burrow through sediments were efficient producers of organic carbon packaged for rapid burial in the form of fecal pellets. Burial of this carbon led to an increase in the oxygen content of the atmosphere (32 grams of O_2 for every 12 grams of carbon buried), and the increased oxygen, decreased carbon dioxide, and lowered temperature caused extinction of large-celled microalgae.

photosynthesis was switched off. RUBISCO-containing cyanobacteria sailed through unscathed because they pump CO_2 into their cells by mechanisms that microalgae do not have.

Perhaps this is what happened on a global scale toward the end of the Precambrian. But perhaps not. There is no good way to measure exactly the quantities of CO_2 and O_2 in the late Precambrian atmosphere, so there is no way to be certain their concentrations were at the levels required to shut off microalgal photosynthesis. And more work still needs to be done on the fossil record to show for sure that big cells were killed off gradually rather than in a single mass extinction at the end of the Precambrian. In the Phanerozoic, environmental change is often associated with extinction. This was true also with the great extinction of the late Precambrian, but details of the role such change played have yet to be worked out.

Prelude to the Phanerozoic

The familiar schoolbook history of life starts with small, shelly fossils that mark the base of the Cambrian System of Phanerozoic-age rocks dated to be about 550 Ma old. This was preceded by the last great geologic event of the Precambrian, the Varangian ice age marked prominently in the rock record by glacial deposits 590 Ma in age and older (see plate 5). Sandwiched between these markers are the oldest many-celled animal fossils, by Phanerozoic standards an odd assemblage of large, diverse, but structurally rather simple, mainly worm-like and jellyfishlike creatures that make up what is known as the Ediacaran Fauna. Their age within the window 590 to 550 Ma ago is not as tightly constrained as one might like, but by the best evidence it seems to be about 560 Ma, much closer to the Cambrian than to the glaciation.

The fossils were found first by Reginald C. Sprigg in March 1946 at an abandonded copper, lead, and zinc mine called Ediacara (an aboriginal term said to mean "foul waters") situated in the Flinders Ranges north of Adelaide in South Australia. To memorialize their source, Sprigg named these saucer-sized, jellyfishlike specimens *Ediacaria flindersi*. Study of the fossils was soon taken up by Martin Glaessner (one of the pioneers we met in chapter 2) and his colleague, Mary Wade, and over the following decades other workers found the same kinds of fossils in rocks of the same age on four other continents. The most important sites in addition to Australia are in Namibia, Newfoundland, England, the Ukraine, and Russia's Ural Mountains and White Sea coast. More than a dozen different genera have been named.

Several books (beginning with Glaessner's 1984 opus, *The Dawn of Animal Life*) and reams of scientific papers have been written about these odd, earliest, many-celled animals. The fossils are wonderfully interesting, discussed at length in practically all first-year biology and

Figure 9.11 Small skeletal fossils (*Chancelloria*) of the kind that mark the beginning of the Cambrian Period (and Phanerozoic Eon) of geologic time dissolved out of 550-Ma-old limestones from Siberia. The intricately sculptured fragments are thought to have fit together to make up the chain mail-like body armor of a primitive sluglike metazoan. (Bar for scale represents 1 mm. Courtesy of S. Bengtson, Swedish Museum of Natural History, Stockholm.)

Figure 9.12 A fossil jellyfish (*Mawsonites*) of the Ediacaran Fauna from the 560-Ma-old Pound Quartzite of South Australia. (Bar for scale represents 1 cm. Courtesy of B. N. Runnegar, University of California, Los Angeles.)

paleontology texts. But their true nature and what they reveal about the origin of animals are controversial. Some experts argue that all but a couple of the fifteen or so Ediacaran genera are not obviously related to later life, whereas others claim that all but two fit in. And while some suggest the fauna was ecologically bizarre—made up of animals that lacked mouths, guts, and anuses and lived on food manufactured by endosymbiotic microorganisms—others view it as a standard mix of heterotrophs like faunas of the Phanerozoic.

What can be said with certainty is that the Ediacaran organisms are (1) early-evolved, many-celled animals, the oldest we now know; (2) soft-bodied, devoid of mineralic hard parts (except probably for the head-shielding carapace of a wormlike form known as *Spriggina*); (3) paralleled in the fossil record by the appearance of tracks, trails, and burrows (trace fossils) that record the rise of mobile muck-processing animals; and (4) either the ancestors of or an evolutionary side

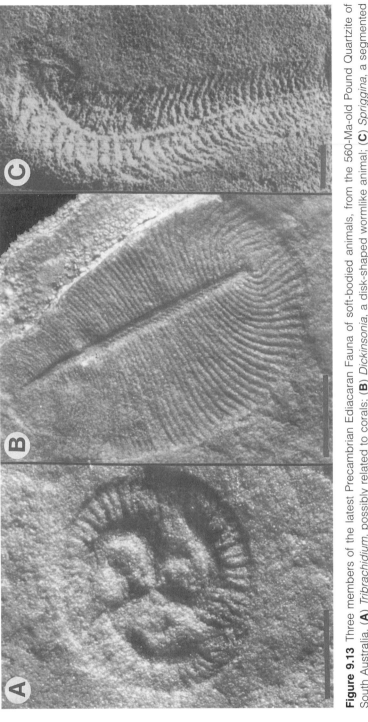

Figure 9.13 Three members of the latest Precambrian Ediacaran Fauna of soft-bodied animals, from the 560-Ma-old Pound Quartzite of South Australia. (**A**) *Tribrachidium*, possibly related to corals; (**B**) *Dickinsonia*, a disk-shaped wormlike animal; (**C**) *Spriggina*, a segmented worm having a distinctive, thickened head shield (*upper right*). (Bars for scale represent 1 cm. Courtesy of B. N. Runnegar, University of California, Los Angeles.)

branch that preceded, the shelly fauna that marks the beginning of the Phanerozoic.

However the Ediacaran animals may have made their living, they would never have evolved had there not been before them an almost endless train of microscopic organisms that invented, in turn, oxygen-producing photosynthesis and oxygen-consuming respiration, the organelle-laden eukaryotic cell, mitotic cloning, meiosis, and sexual reproduction. Evolution builds step by step. We, like all other animals, are a product of that process.

10

Solution to Darwin's Dilemma

The Adventure of Science

Science is the greatest adventure ever devised. The past, present, even the future of life, Earth, and all beyond are within its scope. But the quest has only begun. Modern science is of such recent vintage and the Universe so vast and varied that it will be millenniums and more before the venture comes to a close. How is science done, by whom, and why? The history of Precambrian paleobiology holds useful insights, which we will explore first in this chapter. Then we will take a look at the take-home lessons from the newly gained knowledge of life's early history before briefly considering evolution's most recent great invention, intelligence, which rather remarkably owes its origin to biology first developed billions of years ago.

The Human Side of Science

In the popular media, scientists are often pictured as other-worldly—nerdish, almost robotlike creatures that communicate in polysyllabic arcana and are privy to "hidden knowledge." Sometimes they're shown as benign, even useful in their way, but almost never as the sort of people you'd like to live next door to. This cartoon view is distant from the truth. Actually, of course, scientists have the same strengths, fears, and foibles as everyone, a range spanned by early luminaries in the hunt for ancient life (chapters 1 and 2). Some of these were kind and warm-hearted (such as Boris Timofeev), even courtly and self-effacing (Martin Glaessner); others pugnacious (J. W. Dawson) and perhaps imperious (Preston Cloud). Timidity was not unknown—as when Elso Barghoorn put off for years the irrevocable step of writing his part of the pivotal article on the Gunflint fossils—

nor was stubborn boldness, as when Dawson fought on and on for *Eozoön*. One (Dawson) aspired to disprove evolution and unseat Darwin, and a few sought to one-up the competition by disparaging the work of others (as when Percy Raymond and A. C. Seward belittled C. D. Walcott's prescient finds) or by elbowing their way into someone else's discovery or cheating to protect their turf (as in the Cloud-Barghoorn battle). But like Charles Doolittle Walcott, the founder of the field, all were imaginative, creative, notably intelligent; all impressively hard-working and deeply curious about the natural world; all shared a strong desire to move the science to higher levels of understanding.

Clout in Science

It is no coincidence that practically all the pioneers belonged to the power elite, a status that upped the odds for acceptance of their views. Three traits contributed to their clout: professional expertise, ties to prestigious institutions, and the political "correctness" of the views espoused.

Some of the players, mostly those in secondary roles, entered the scene as acknowledged experts—William Carpenter and Karl Möbius, the micropaleontologists caught up in the *Eozoön* quarrel; A. C. Seward, the paleobotanist who took issue with Walcott's views of ancient life; Stanley Tyler, the geologist-mineralogist who unearthed the famous Gunflint fossils. But before the discoveries of the mid-1960s there was no defined field to be schooled in, so the leading movers and shakers were all trained in other disciplines and self-taught in this one. Several had towering reputations, most notably Dawson and Walcott, gained by appointment to high positions and friendships with the powerful; and the bona fides of others were affirmed by the eminence of their institutions—Barghoorn by Harvard, Cloud by his Yale degree and ties to the U.S. Geological Survey.

But the social climate and politics also shaped the picture. When Darwin raised the question of the "missing" early fossil record, he was listened to with care since he was speaking from the hub of the scientific world, England and particularly London. This centrality is illustrated by the forum picked for debate of Dawson's *Eozoön*—the Geological Society of London rather than the Canadian hinterlands where it had been found—and the respect paid Carpenter's support of its supposed biologic origin, an appraisal considered certified by his position at the British Museum. And though in a perfect world political correctness would play no role in science, it sometimes does since scientists, like everyone else, can succumb to real-world pressures. It's easy to see now that when Dawson championed *Eozoön* he was acting as a protector of the faith, shoehorning his views to fit his Calvinist religion. The ideas of A. I. Oparin (and J. D. Bernal) on the origin of life meshed with Marxist dialectic materialism, so in the Soviet Union they were "correct." Yet for the same reason, and due to Oparin's friendship with the Stalinist geneticist T. D. Lysenko, the ideas were politically suspect in the West (chapter 4). Furthermore, the miscues of the Russian Precambrian pioneer B. V. Timofeev might have been avoided had he not been cut off by politics from the non-Soviet world.

Though science strides on feet of facts, its march is sometimes hastened if the "right view" is pushed by those who have the "right credentials" and "right connections."

How Is Science Done?

Of various styles of science, the most common is "Science by Conventional Wisdom," where findings are deemed valid if they agree with what is "known." The historian and philosopher of science Thomas Kuhn called this process "normal science" where the aim is to discover facts that fit a view already universally held (in Kuhnian-speak, to "articulate the paradigm"). Most good scientists are masters of the game, and in fields that are mature, based on bodies of knowledge amassed over many years, this style works well because what "everyone knows" *is* usually right. But because it can work only if the basic questions have been answered and a solid foundation has been laid, it failed miserably in the start-up years of Precambrian paleobiology when most crucial questions hadn't even been asked. Had conventional wisdom won out, A. C. Seward would have prevailed, no one would even have tried to hunt for ancient microbes, and life of the Precambrian would be unknown and still thought unknowable.

A style closely related is "Science by Authoritative Assertion," where views are certified by what the experts say. This, too, failed as shown by Dawson (*the* expert on Canadian geology), Carpenter (*the* leading micropaleontologist in Britian), and Seward (*the* foremost paleobotanist worldwide). Some authorities (such as the micropaleontologist Möbius and the geologist Tyler) hit the nail on the head, and in fields that have proven track records, the experts *are* very often right. But expert opinion is wedded so firmly to views assumed to be correct that it can be more hindrance than help in the lift-off phase of a brand new discipline.

A third style is "Science by Smart Luck." Unlike the unwitting "dumb luck" of winning a lottery or hitting the jackpot on a slot machine, Smart Luck depends on skill and training. It paid dividends as the field emerged because the early workers were primed to capitalize on what turned out to be lucky breaks. A neophyte USGS geologist, Walcott was fortunate to be assigned the hunt for Precambrian life—but was equipped to cash in by his apprenticeship with state geologist James Hall and his firsthand knowledge of *Cryptozoon*. By chance Oparin heard Timiryazev lecture on Darwinism—but was prepared to place animal-like heterotrophs at the base of the Tree of Life by studies of botany begun in his youth. Tyler was an expert on minerals, not ancient microbes, but was prepared to see the Gunflint objects as possibly fossil by his one formal course in paleontology. And because of a wartime stint studying microscopic fungi in Panama, Barghoorn was in a position to confirm Tyler's guess.

If the right thing happens to the right person at the right time, Science by Smart Luck works. But serendipity is all too unpredictable. This is not a strategy one should rely on.

"You Must Not Fool Yourself"

In 1974 the brilliant CalTech physicist and teacher Richard Feynman gave an address to his university's graduating class in which he unveiled his First Principle of Science: "You must not fool yourself—and you are the easiest person [for you] to fool!" As Feynman well knew, any scientist would be elated to make a breakthrough discovery, to unearth the Rosetta Stone of his or her field. For most, the motivation is not money, not fame, not the short-lived glory that comes with a major new find—it is to make a difference to human knowledge, to contribute new and lasting insight about things not understood before. And therein lies the rub, for so strong is the desire to make a breakthrough that it's easy to become smitten by a notion that is later shown to be dead wrong. Recent examples include polywater, cold fusion, and, as we'll see in the epilogue to this book, perhaps even the claims of ancient life on Mars.

How does science guard against such errors? There is one style that can always be counted on, "Science by Facts." Here claims are accepted only if they are thoroughly tested, debated, and validated by the extended scientific community. At its core, good science is studied skepticism. All conjectures must pass muster, and the more they depart from accepted views—the more at odds they are with conventional wisdom—the more rigorous the testing must be. The late Carl Sagan said it well: "Extraordinary claims require extraordinary evidence."

When first announced, Darwin's claim of evolution was not only extraordinary but deeply unsettling. Scientists, theologians, philosophers, even laypersons voiced opinions as the theory was examined down to its smallest detail. It actually was so threatening to conventional wisdom that for years after publication of *On the Origin of Species* formal public debates ("Science Lectures for the People") were convened, often with the case for evolution argued by T. H. Huxley (1825–1895), the English biologist known widely (if not always fondly) as "Darwin's Bulldog." Hurdles were high, acceptance was grudging, even among scientists. As Huxley put it, thinking back to when *On the Origin of Species* appeared: "There is not the slightest doubt that, if a general council of the church scientific had been held at the time, we should have been condemned by an overwhelming majority."

The skepticism was deserved, for Darwin's idea flew in the face of accepted "truth." Yet as the theory withstood this trial by fire, its depth and strength came to be obvious even to most skeptics, and by now it is so firmly fixed that it stands as the rock solid foundation—the GUT, the Grand Unifying Theory—of all life science. Dawson's claim for *Eozoön* was also extraordinary, but science's skepticism showed it not to be credible. Seward's qualms about Walcott's "Precambrian bacteria" were in the same tradition, and though he guessed wrong, his skepticism was on the mark—Walcott had only flimsy scientific evidence to support his claim. And those questioning the

Gunflint and Bitter Springs fossils in the late-1960s were right as well. Thanks to their doubts, we now have hard-nosed rules to separate the serious from the spurious, the bona fide from the bogus.

Why did Darwin's evolution prevail, and in the contest to unearth ancient life, why did Dawson lose, Seward and Walcott tie, and workers of the 1960s score the winning goals? Facts. Darwin won because *The Origin* is massed with facts that support his thesis. Dawson lost because his claim was shown to be at odds with the facts. Seward and Walcott tied, since neither had the facts to prove his case. And the leaders of the 1960s—Tyler, Barghoorn, Cloud, Timofeev, Glaessner— won big because their facts tally with all that has been learned in the decades since. In mature fields of science, Conventional Wisdom and Authoritative Assertion are right more often than wrong. And even in an emergent field, Smart Luck can come in handy. But these styles can't be counted on. At the end of the day, there is only one style that is always reliable: Science by Facts. Hardened by the fires of studied skepticism, firm facts—backed by a cascade of their natural consequences, then theirs, and theirs—always win.

Take-Home Lessons

Since the mid-1960s a torrent of discoveries about life's earliest history has flooded the field. New findings prompt better questions, and as shown even by the greatly simplified overview presented in this book we can now see much more clearly than 30 years ago the prime problems to be solved. Two breakthrough innovations were pivotal in early evolution, oxygen-producing (cyanobacterial) photosynthesis and eukaryotic sex; but the advent of neither is dated precisely, and there are many other unknowns as well. Table 10.1 is a report card on the status of the field, a point-counterpoint summary of what is known and what is not about six highlights in life's early development.

Solution to Darwin's Dilemma

In *On The Origin of Species*, in 1859, Darwin posed the dilemma: "If the theory [of evolution] be true . . . the [Precambrian] world swarmed with living creatures. [But] to the question why we do not

Table 10.1 Highlights in Life's Early History

What Is Known	Unsolved Problems
The Beginnings of Life	
1A. Life's origin followed a simple path—CHON begat monomers, monomers begat polymers, polymers begat cells and populations (chapters 4 and 5).	**1B.** There seems to be good understanding of monomer formation, less of the way polymers formed, only hints how cells came to be (chapters 4 and 5).
2A. Life began very early in Earth history to include by 3,500 Ma ago a flourishing microbial zoo of nonnucleated microorganisms (chapter 3).	**2B.** The beginnings of evolution raise questions that may never be firmly answered since Earth's earliest rock record is lost forever (chapter 6).
Life's First Great Success	
3A. Today's ecosystem of eaters and eatees, anaerobes and aerobes, was invented by microbes long before the advent of plants and animals (chapter 5).	**3B.** The crowning innovation in early history was oxygenic photosynthesis, possibly but not certainly as early as 3,500 Ma ago (chapters 3 and 5).
4A. Cyanobacteria came to be kings of the planet, living almost everywhere and giving off oxygen that dramatically changed the world (chapters 6, 7, and 8).	**4B.** The fossil record is too sparse to show whether other microbes also were slow-evolving (hypobradytelic) ecologic generalists (chapter 8).
Life's Second Great Invention	
5A. Prepackaged evolution gave genesis to chloroplast- and mitochondria-containing nucleated cells that multiplied by mitotic cloning (chapters 5 and 9).	**5B.** Exactly when is unknown and the early fossil record of planktonic microalgae (sphaeromorph acritarchs) is woefully incomplete (chaper 9).
6A. Sex was evolution's second great innovation because it spurred the rise of plants and animals having new specialized ways to live (chapter 9).	**6B.** Sex only possibly dates from 1,100 Ma ago, and the path to large many-celled forms by 600 Ma ago is seen only dimly (chapter 9).

find rich fossiliferous deposits belonging to these assumed earliest periods . . . I can give no satisfactory answer. The case at present must remain inexplicable."

Though the puzzle of the "missing" early fossil record lived on for more than a hundred years, its solution is now so obvious as to be mundane. The Precambrian world did *indeed* swarm with living creatures, but until near the close of this vast eon these were microbes and microalgal cells so tiny and fragile that they would never have been unearthed by conventional fossil hunting. As Preston Cloud phrased it, writing of his early failures in the search for ancient life, he, like all others, "had been looking for the wrong things in the wrong rock types by the wrong methods." As we've seen in the chapters of this book, the solution to Darwin's Dilemma came in fits and starts over a long, error-prone century to be. It was spurred finally in the 1960s by use of new methods to answer new questions advanced by a new field of science, Precambrian paleobiology.

Evolution Evolved!

One take-home lesson from the past three decades of discovery stands above all others: the rules of Precambrian evolution differed decisively from those of the Phanerozoic. Evolution evolved!

Though evolution's evolution is a fundamental fact of the history of life, its discovery was unforeseen. Everyone had expected the standard rules of the Phanerozoic to apply equally to the Precambrian. Early organisms would be smaller, simpler, perhaps less varied, but they were universally thought to have evolved in the same way and at the same pace as later life. That this turned out not to be true shows the beauty of how science works. Even the strongest hunches of learned experts can never take the place of facts. Nature is chock-full of surprises. The British biologist J.B.S. Haldane put it well: "The Universe is not only queerer than we suppose; it is queerer than we *can* suppose."

The pivotal point in the evolution of evolution was the advent of sex in eukaryotes perhaps 1,100 Ma ago. Sex not only increased variation within species, diversity among species, and the speed of evolution and genesis of new species (as we saw in chapter 9), but it also brought the rise of organisms specialized for particular habitats. By early in the Phanerozoic, the mitotic (body cell division) part of plant life cycles had come to focus on vegetative growth, giving rise as the eon unfolded to trees, shrubs, and grassy vegetation honed to specific settings. At the same time, the meiotic part developed specializations for increasingly reliable reproduction as the flora evolved from spore producers to wind-pollinated naked-seed plants (gymnosperms) and eventually to flowering plants (angiosperms), where propagation is often aided by pollen-carrying insects. Animal evolution followed a parallel course. In adults (the mitotic part of the life cycle), changes in limbs and teeth led to increasingly specialized ways of feeding, while the meiotic part evolved improved means of reproduction—from the broadcast larvae of marine invertebrates and roe of fish to the soft egg masses of amphibians, the hard-shelled eggs of reptiles and birds, and the protected embryos of mammals.

Phanerozoic plants and animals were large, many-celled, and specialized for particular settings. Most species were relatively short lived (horotelic), and the eon was punctuated by many episodes of extinction, each followed by the rise and diversification (adaptive radiation) of organisms that survived. But the earlier and much longer Precambrian was different. Almost all species were microscopic, nonsexual, and prokaryotic. Major evolutionary changes were intracellular and biochemical. The most successful Precambrian microbes were cyanobacteria, exceedingly long-lived (hypobradytelic) generalists able to tolerate a wide range of environments. Mass extinctions were rare and mostly limited to big-celled microalgal eukaryotes.

The take-home lesson is clear: life's history is divided into two great epochs, each having its own biology, style, and tempo—the Precambrian "Age of Microscopic Life," a world of microbial long-lived ecologic generalists; and the Phanerozoic "Age of Evident Life," ruled by eukaryotic short-lived specialists. That evolution itself evolved is a new fundamental insight that, together with the astonishing antiquity of living systems (chapter 3), stands out as one of the most striking findings to have come to light about life's long history since Darwin first showed how evolution works.

Déjà vu All Over Again

The two-part history of life is rather like what happened when early settlers spread to the interior of North America. The analogy is rough,

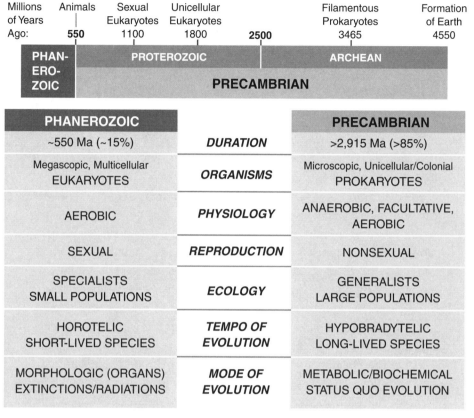

Millions of Years Ago:	Animals	Sexual Eukaryotes	Unicellular Eukaryotes		Filamentous Prokaryotes	Formation of Earth
	550	1100	1800	**2500**	3465	4550

PHAN-ERO-ZOIC	PROTEROZOIC		ARCHEAN
	PRECAMBRIAN		

PHANEROZOIC		PRECAMBRIAN
~550 Ma (~15%)	*DURATION*	>2,915 Ma (>85%)
Megascopic, Multicellular EUKARYOTES	*ORGANISMS*	Microscopic, Unicellular/Colonial PROKARYOTES
AEROBIC	*PHYSIOLOGY*	ANAEROBIC, FACULTATIVE, AEROBIC
SEXUAL	*REPRODUCTION*	NONSEXUAL
SPECIALISTS SMALL POPULATIONS	*ECOLOGY*	GENERALISTS LARGE POPULATIONS
HOROTELIC SHORT-LIVED SPECIES	*TEMPO OF EVOLUTION*	HYPOBRADYTELIC LONG-LIVED SPECIES
MORPHOLOGIC (ORGANS) EXTINCTIONS/RADIATIONS	*MODE OF EVOLUTION*	METABOLIC/BIOCHEMICAL STATUS QUO EVOLUTION

Figure 10.1 The prime take-home lesson: the rules of evolution changed from the Precambrian to the Phanerozoic.

but in interesting ways American history appears to parallel the pale-ontologic past—in baseball-great Yogi Berra's phrase, "It's *déjà vu* all over again!"

The first waves of settlers streaming West included few true crafts-men, specialists in building houses, making crockery, weaving. But needs then were the same as now—a place to live, plates to eat from, clothes to wear. So, like generalists of the Precambrian, the most suc-cessful pioneers were jacks-of-all-trades able to cope in a variety of settings. As settlements grew, so did specialization—with the arrival of merchants, tradesmen, and artisans. A new breed of house builders could build, but not make plates or clothes. Clothiers could clothe, but not erect a house. Products were made better and faster, and as in the history of Phanerozoic life specialization led to diversity as crafts spawned new subcrafts—roofers and plumbers to assist the house builders, button makers to supply the clothiers. But also as in the Phanerozoic, specialization upped the odds of extinction. Candle makers lost out when candles were displaced by gas lamps; gas-lamplighters, when electricity became commonplace; blacksmiths, when automobiles took the place of horse-drawn buggies. Specialization and short-lived success go hand in hand. Diversity abounds but the varied components are ever-changing.

Specialization also breeds interdependence, and this, too, spurs extinction. Even today the rise of computers has not only put typewriter manufacturers out of business but people dependent on them as well—secretary-typists, typewriter repair teams, suppliers of parts, and those overseeing and teaching these specialties. The biology of ecosystems is interdependent in much the same way, so it is not surprising that a similar rippling of cause-and-effect happened many times during the Phanerozoic, when extinction wiped out key phytoplankton at the base of the food web.

As in the Phanerozoic, specialists predominate in our modern, citified societies, whereas jacks-of-all-trades—a local handyman, for instance—are hard to find. We accept this as "progress," though perhaps we shouldn't be so eager. The next time your car won't start, or your phone is out of order, or your computer crashes, remember the early settlers—like generalist cyanobacteria of the Precambrian, they would be able to cope!

The Roots of Human Intelligence

In addition to increased specialization there is at least one other trait that sets life of the Phanerozoic apart from that of the Precambrian: intelligence. But it, too, has Precambrian roots. How could this be, and how far back do the roots extend?

In some circles, intelligence is claimed to be exclusively human. Almost certainly this notion is wrong. What about chimpanzees and dolphins, or dogs, cats, and horses? They all seem to think. Even pigs are said to be surprisingly smart. Consider a bird of prey, an eagle, for instance. It doesn't dive to where its victim *was* but to where it *will be* when the eagle swoops to snatch it. Doesn't that require forethought? Even a frog lying quietly in wait for an insect meal, or a fiddler crab waggling its claw to attract mates and stake out its territory, seems to know what *is* and foresees what *will be* going on around it. How much of this reflects intelligence?

Definitions of intelligence vary. But most agree that the trait is shown by the way an organism interacts with and affects its environment, and that it is rooted somewhere deep in the Tree of Life. Why is it that some organisms are smart and others evidently not? Microbes, protists, fungi, and plants clearly do not qualify. And there's no reason to believe that animals such as sponges, corals, or sea lilies are particularly smart, either—all they seem to do is stay glued in place and "vegetate." Perhaps that's the answer. Trees, shrubs, or the builders of a coral reef don't move about, so they are unlikely to need high-quality thinking. They require CHON and energy, of course, but these needs are met easily by a stay-put lifestyle in which they simply soak up sunlight or tiny food particles that settle into their environs.

But to move from one place to another, *mobile* animals need large amounts of energy and consequently much more nourishment than sessile forms of life. Since no organism with such demands can obtain CHON reliably without at least a semismart strategy for foraging, the roots of intelligent behavior likely lie in the origin of feeding strategies in primitive, mobile eukaryotic heterotrophs.

Among the most successful early-evolved many-celled animals were those best able to ferret out foodstuffs. They had food-seeking and food-engulfing apparatus localized at their front "encounter" end, defining a front-to-rear bodily organization carried over to all descendants, including humans. Over time, the nerve systems to drive the food-ferreting apparatus also came to be localized at the encounter end. This made for shorter nerve routes and better efficiency, but concentration of these delicate systems in a single area made them more vulnerable to life-threatening damage such as being tossed about in storm-driven waves or squashed by a tumbling rock. This serious drawback was offset by the development of forms having a protective thick skin (carapace) overlying the nerve-rich, vulnerable area. The success of these animals led to increased cephalization—the development of a helmeted head region to protect the neural equip-

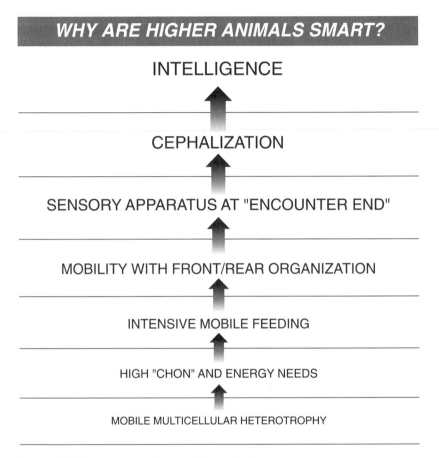

WHY ARE HIGHER ANIMALS SMART?

INTELLIGENCE

↑

CEPHALIZATION

↑

SENSORY APPARATUS AT "ENCOUNTER END"

↑

MOBILITY WITH FRONT/REAR ORGANIZATION

↑

INTENSIVE MOBILE FEEDING

↑

HIGH "CHON" AND ENERGY NEEDS

↑

MOBILE MULTICELLULAR HETEROTROPHY

Figure 10.2 Major steps in the evolution of intelligence in animals.

ment used to receive and process sensory information. From these beginnings evolved an increasingly intricate sensory apparatus and associated nerve systems, giving rise eventually to forms we would call truly intelligent, able to interact with and affect their environment in complicated ways.

Viewed in this manner, intelligence is a logical outcome of the rise of mobile, many-celled animals. But mobile heterotrophy itself arose to answer the necessities of life—the needs for CHON and energy. And these needs as well as the two strategies evolved to meet them, autotrophy and heterotrophy, are products of much earlier evolution, set in motion billions of years ago when the planet and life had just been born. The seeds that ultimately flowered in human intelligence can be found in the ecologic structure of eaters and eatees that dates from Earth's earliest history.

Our cells, the chemistry that keeps us alive, the way the organs in our bodies are arranged, how we grow and reproduce—*everything* that makes us the way we are—house memories of an unimaginably long evolutionary journey that began billions of years ago. From microbes to man, from ignorance to intelligence, the story of life's evolution from its primal Precambrian past holds the key to understanding why we are and where we have come from.

EPILOGUE
EXTRAORDINARY CLAIMS!
EXTRAORDINARY EVIDENCE?

Fossils, Foibles, and Frauds

Extraordinary claims require extraordinary evidence.
Carl Sagan, on numerous occasions

The Goal Is to "Get It Right"

From time to time, mind-boggling scientific breakthroughs are trumpeted in the press: chemicals to cure cancer or control AIDS, the cloning of Dolly the sheep, life that thrives above the boiling point of water, dormant seeds revived after a thousand years, evidence of past life on Mars. Because there are more scientists than ever before, breakthroughs come at a quickening pace. Yet truly major ones are still few and far between, precious rarities worth savoring. And though one may think they are conjured out of thin air—the discoverer shouting "Eureka!" and jumping for joy—practically all have decades-long gestation periods. Discovery of Earth's earliest fossils is a case in point, a claim of extraordinary antiquity for life that comes on the heels of 30 years of finding vast, ancient fossil records, giving the claim an extraordinarily firm backing.

The goal in science always, of course, is to "get it right." But getting a wrong answer matters more at some times than at others. Science would not be set on its ear if a species of trilobite claimed to be "new to science" were, years later, shown to be identical to one of scores from the same rock layer. But it would be a disaster if science accepted the "breakthrough discovery" of a trilobite wrongly reported from a bed a billion years in age, nearly half a billion years older than any animal ever found. The first error would be minor, unlikely to have lasting impact. The second would be serious, because its acceptance would tear down a cornerstone of what we think we understand. Because a claim of a breakthrough is just that—the announcement of a find said to be surpassingly important—those making the claim have special responsibilities. The late Carl Sagan was right: extraordinary claims *do* require extraordinary evidence.

Extraordinary claims are far older than science itself, attention-grabbers mastered long ago by witch doctors and tribal medicine men, soothsaying astrologers and court magicians, the "savants" of their time. Claims shown to be correct are well chronicled, such as those of Nobel laureates. But equally fascinating are claims doubted but not yet sorted out (as we will see in chapter 12), and claims that failed, a side of science seldom seen—of error, despair, sometimes even of eventual redemption of the claimant if not the claim.

This two-chapter epilogue deals first, in this chapter, with two claims that fizzled, both dating from the early 1700s. The first was the discovery of the skeleton of a human said to have drowned in Noah's Flood and for many decades thought to be proof of Biblical Truth, until finally it was shown to be a misidentified huge fossil salamander. The second was the reported find of unbelievably "perfect" fossils of butterflies, bees (one with its honeycomb), birds (with freshly laid eggs), spiders (some with webs, one devouring a fly), and even "fossilized imprints" of the Moon and stars—fanciful fake fossils carved, buried, and then "dug up" to mystify and embarrass an arrogant professor. The next chapter explores a claim so recent that it is not yet resolved—the extraordinary claim of past life on Mars that, if it proves to be true, would rank among the most momentous discoveries of this or any other century. Though these tales are entertaining and fun to think about, their true value lies in what they show about how far science has advanced over the last 250 years and what it takes for a "major breakthough" to stand the test of time.

"Man, a Witness of the Deluge"

One of the most famous extraordinary claims in paleontology comes from the early eighteenth century and the studies of Dr. Johann Jacob Scheuchzer (1672–1733), a Swiss physician and naturalist of broad-ranging interests. An outstanding academic and best known for his

Figure 11.1 Johann Jacob Scheuchzer. (By permission of the Linnean Society of London.)

studies of geology and paleontology, he was a respected scholar also of mathematics, geodesy, geography, literature, and numismatics (coin collecting). The *English Cyclopedia* of 1856, more than a century after his death, acclaims him as a savant of "indefatigable industry and extensive knowledge."

Scheuchzer was well schooled in the Christian tradition. Like many learned men of the day he was a confirmed Diluvialist, and his view of Earth history was molded mightily by the story of Noah and the Flood. In his 1709 treatise *Herbarium Diluvianum*, for instance, he not only pegs the Great Deluge as the agent that formed fossil-bearing rocks, but he actually dates the Flood as happening in the spring, probably in May, because of the "tender, young, vernal" state of entombed seed cones. (This notion—likely borrowed from Englishman John Woodward's 1695 *Essay towards a Natural History of the Earth* —was hotly disputed by those who argued that "ripe" fruits in the very same deposits proved that the Deluge happened in autumn.)

But Johann Jacob Scheuchzer is most remembered for a later contribution, a discovery that was to be the center of a debate for nearly 100 years. In 1725 Scheuchzer uncovered the partial skeleton of a large, elongate, obviously vertebrate animal in limestone quarried near Oeningen, Baden, Germany. To Scheuchzer the preserved skull and backbone looked decidedly humanlike and totally different from any fossil he had ever seen. There could be only one explanation— this must be the remains of a man drowned in the Flood, the bones of one of those miscreants whose sinful ways brought upon the world the catastrophe of forty days and nights of the Great Deluge! Elated with this proof of Biblical Truth, Scheuchzer made it the keystone of his monumental *Physica Sacra* published five years later. In his words:

> It is certain that this . . . is the half, or nearly so, of the skeleton of a man: that the substance even of the bones, and, what is more, of the flesh and of parts still softer than the flesh, are there incorporated in the stone. We see there the remains of the brain . . . of the roots of the nose . . . and some vestiges of the liver. In a word it is one of the rarest relics which we have of that cursed race which was buried under the waters [of the Noachian flood].

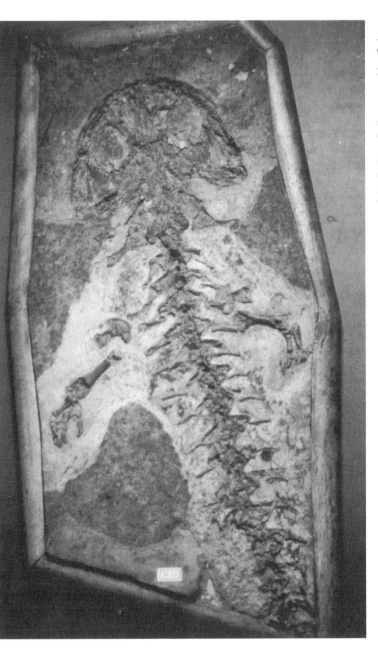

Figure 11.2 The original (holotype) specimen of *Homo diluvii testis*, "Man a witness of the Deluge," discovered by Scheuchzer in 1725 and thought by him to be the flattened skull and backbone of a man drowned in Noah's Flood. Known now to be the upper torso of a giant salamander fossilized in 15-Ma-old (Late Miocene) freshwater limestone, the skeleton was hailed for nearly 100 years as direct proof of Biblical Truth. (Teyler Museum specimen 8432, Case V29 in the second room of the Minerals and Fossils Cabinet.)

To memorialize the find, Scheuchzer dubbed the specimen *Homo diluvii testis*—literally, "Man, a witness of the Deluge"—and for dramatic effect included a moralistic couplet penned by church deacon Miller von Leipheim:

> Betrübtes Beingerüst von einem alten Sünder,
> Erweiche, Stein, das Herz der neuen Bosheitskinder.

In English (as translated by Herbert Wendt):

> Afflicted skeleton of old, doomed to damnation,
> Soften, thou stone, the heart of this wicked generation!

Within only a few years, Scheuchzer's antediluvial man was hailed throughout Christendom as irrefutable evidence of the Holy Word.

But not all scholars were convinced. Among the first to question Scheuchzer's claim was another Swiss physician and naturalist, Johannes Gessner (1709–1790), Scheuchzer's former student and his successor to the professorship at the Carolinum in Zurich. Gessner raised many an eyebrow by suggesting in 1758 that Scheuchzer's famed find might actually be bones of a large though only partially preserved fossil fish. However, the doubts of Gessner and a few others notwithstanding, to most the matter was settled. The Bible was beyond question. Scheuchzer's view was eminently sensible, a confirmation of what conventional wisdom already knew.

At his death in 1733, Scheuchzer bequeathed the specimen to his son, who passed it along to *his* son. In 1802, now world-renowned, *Homo diluvii testis* was purchased from Scheuchzer's grandson by the Teyler Museum at Haarlem, just west of Amsterdam, where it is housed to this day.

Cuvier "Cleans" the Specimen

A few years later the famous fossil came under the scrutiny of Baron Georges Cuvier (1769–1832), the French comparative anatomist credited as a founder of the science of paleontology. In 1810, with the annexation of North Germany and the entire kingdom of Holland, Napoleon Bonaparte's empire reached its widest extension. Cuvier was Napoleon's minister of education, and in 1811 he journeyed to Amsterdam as head of a commission to review and improve the

Figure 11.3 Johannes Gessner. (By permission of the Linnean Society of London.)

Figure 11.4 Teyler Museum, in Haarlem on the outskirts of Amsterdam, built in 1780 behind the home of the museum's benefactor, Pieter Teyler van der Hulst, and the oldest museum built as such in the Netherlands. A childless widower and owner of a prosperous silk mill, at his death in 1778 Teyler bequeathed his house and much of his fortune to establish a foundation bearing his name for "promoting [the Christian] Religion, encouraging the Arts and Sciences, and for the Public Benefit." The museum consists of four "cabinets" (departments), of Art, Coins, Physics, and Minerals and Fossils.

Figure 11.5 Baron Georges Cuvier. (By permission of the Linnean Society of London.)

Dutch educational system. He had written ahead to the directors of the Teyler Foundation requesting permission to examine and "clean" the specimen, to chip away rock that masked parts of the skeleton from Scheuchzer's view. *Homo diluvii testis* was far and away the most precious specimen of the museum's collections—kept under lock and key and cleaned not even by Scheuchzer when it was unearthed in the 1720s—but to curry favor with the occupying French, the directors granted Cuvier's wish.

With painstaking skill, Cuvier uncovered the fossil's "arms" and "hands." Rather than being of human origin, as Scheuchzer had surmised, these turned out to be the short forelimbs and clawed fore-feet of a salamanderlike amphibian. Cuvier was not surprised. On the basis of Scheuchzer's published drawing, Cuvier had long suspected the fossil to be a large salamander. Guided by a sketch of a sala-mander skeleton brought with him from France, he chipped away just those parts of the rock needed to prove his point. If one compares the actual specimen with the drawings of Scheuchzer and Cuvier, it is easy even today to discern Cuvier's hand, to see exactly where he "cleaned" the storied skeleton so many years ago. Cuvier showed *Homo diluvii testis* to be a fossil salamander belonging to the genus *Andrias*, so in 1831 it was renamed *Andrias scheuchzeri*, in honor of Scheuchzer.

But the salamander's large size remained a puzzle. Its upper torso, the part preserved in Scheuchzer's specimen, is nearly half a meter long, so the animal would have been more than a meter from nose to tail, much larger than any salamander ever seen. This quandary, too, was soon resolved. In 1829, nearly two decades after Cuvier's restudy of the specimen and more than a century after it was found, the Ger-man explorer Philipp Franz von Siebold (1796–1866) discovered gi-ant batrachian salamanders living in southern islands of the Japanese archipelago, modern forms so similar to the fossil in size and bone structure that they have been named a subspecies, *Andrias scheu-chzeri japonicus*. In the century and a half since, thousands of speci-mens of fossil plants and animals, including hundreds of different species, have been unearthed from the freshwater limestone at the Oeningen quarry, known now to date from the Late Miocene (about 15 Ma ago). Though rare, giant salamanders are known from some twenty-six examples, several nearly a meter and a half long.

For nearly 100 years after its discovery in the early 1700s, *Homo diluvii testis* was acclaimed as proof of Noah's Flood. But it was no proof at all, simply a misinterpreted fossil salamander, unusually large and new to science. Still, the name of Johann Jacob Scheuchzer is

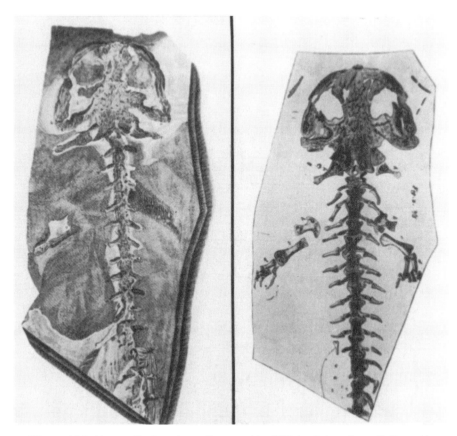

Figure 11.6 *Homo diluvii testis*, as illustrated in 1735 by Scheuchzer (*left*), and in 1824 by Cuvier (*right*) after Cuvier had "cleaned" the fossil (renamed *Andrias scheuchzeri* in 1831) to expose its forelimbs and clawed forefeet. The cleaning showed it to be a salamander, not, as Scheuchzer thought, a human skeleton buried in the Biblical Flood.

forever linked to his mistakenly identified "Man, a witness of the Deluge."

Beringer's Lying Stones

The legend of "The Lying Stones of Dr. Beringer" tells the story of yet another extraordinary paleontologic claim. Told, retold, and embellished in paleontology courses worldwide for more than two and a half centuries, it is an incredible story of an infamous hoax that provides a rare peek into the workings of natural science in the early 1700s.

The standard story is that of an imperious professor, J.B.A. Beringer, who was duped by his students when they carved and hid fake fossils where the prof was sure to find them. Indeed he did. In the belief that he had chanced on a goldmine, with grand flair he wrote up the discovery in a magnificent opus. But then he was astonished to find his own name spelled out on one of the rocks in Hebrew letters. He was the victim of a hoax! Completely humiliated, he bought back the published volumes and died soon thereafter—penniless, friendless, in deep despair.

A wonderful fable, but only partly true. Some years ago, the real story came to light, thanks to 200-year-old court documents discovered in the Würzburg State Archives. These were the findings:

- In 1726, Johann Bartholomew Adam Beringer (1667–1740) *did* author the grand opus (the *Lithographiae Wirceburgensis*), his third published work. But he did not die until 14 years later and during the interim published two more volumes, one a lengthy treatise on the spread and treatment of the dreaded plague.

- Beringer *was* a professor at the University of Würzburg, Germany. A physician, son of a professor, and recipient of Ph.D. and M.D. degrees, he held the titles of Senior Professor and Dean of the Faculty of Medicine at the University, Chief Physician at Würzburg's Juliusspital (Julius Hospital), and Advisor and Chief Physician to Christoph Franz von Hutten, the "learned and modest" Prince-Bishop of Würzburg and Duke of Franconia (though, strangely for a person of Beringer's status, no portraits of him exist today).

- His treatise *does* depict an astounding zoo of fake fossils. Butterflies, beetles, bees (one together with its honeycomb). Birds (some feathered, others denuded, one in flight, two alongside freshly laid eggs). Flowers (with their stems, roots, and leaves, one being pollinated by a hovering bee). Fish, frogs, salamanders, lobsters. Crabs, millipedes, scorpions, earthworms. Ants and wasps with their nests, spiders (two with their webs, one devouring a fly). Caricatures even of

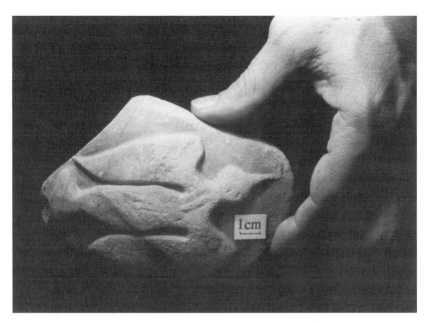

Figure 11.7 One of the surviving specimens of Beringer's famous lying stones, showing a bird in flight. (From the collections of the Geologisch-Paläontologisches Institut der Universität Würzburg, Germany.)

the Moon, Sun, stars, and comets. (Many of Beringer's stones still exist, some on view at the Geologische-Paläontologisches Institut der Universität Würzburg, others at natural history museums in Berlin and London.)

- The famous *Lügensteine* (literally, lying stones) *had* been carved, hidden (on a low hilltop, Mount Eivelstadt, on the outskirts of Würzburg), then "dug up" as part of an elaborate hoax. But the finds were made by diggers employed by Beringer, not by Beringer himself, and the perpetrators were not Beringer's students. Instead, the scam was pulled off by two academics, Herr J. Ignatz Roderick (Professor of Geography, Algebra, and Analysis) and the Honorable [sic] Georg von Eckhart (Privy Councilor and Librarian to the Court and to the University), aided by a German nobleman named Baron von Hof (possibly the backroom financier of the conspiracy but about whom little is known except that he was trundled about in a sedan chair).

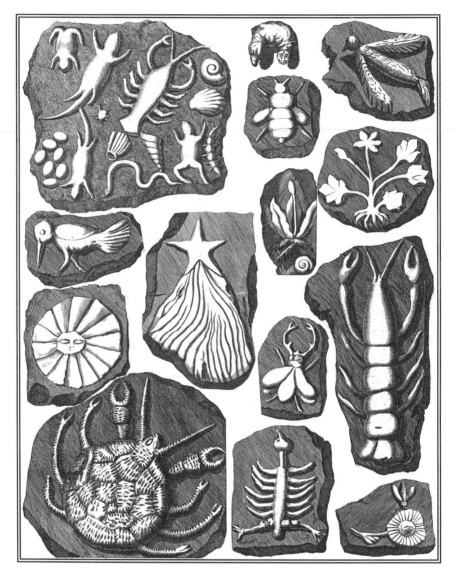

Figure 11.8 Representative figures from the *Lithographiae Wirceburgensis*. Among the various fantastic "oryctics" (things dug from the earth), note the bony bird (*lower center*); a flower being visited by a bee (*lower right*); a smiling Sun (*left center*) and fiery comet (*center*); a plant with root, stem, leaves, and underlying snail (*right center*); a feathered bird in flight (*upper right*); a smiling, two-headed, lizardlike oddity (*upper right center*); and a diverse assemblage of "fossils" (*upper left*).

Figure 11.9 Representative figures from the *Lithographiae Wirceburgensis*. Especially notable are the "perfectly preserved" millipede (*lowermost right*); a spider with its web (*lower right center*); a mermaidlike sea creature (*lower right*); a four-toed lizardlike form (*center right*); a spider devouring a fly (*uppermost left*); ants and wasps with their nests (*upper left center*); and copulating frogs (*lower center*), salamanders (*uppermost right*), and insects (*center and upper left*). The fabricators of Beringer's famous Lügensteine ("lying stones") must have had a sense of humor!

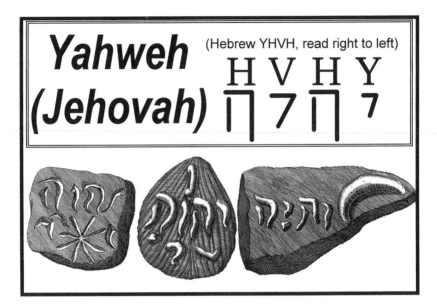

Figure 11.10 Three of the stones depicted in Beringer's *Lithographiae Wirceburgensis* that show the surface marks (accompanied by ornamentation) he interpreted correctly as Hebraic lettering for "YHVH"—Yahweh (Jehovah). Other stones said by Beringer to show the name of God in Latin and Arabic characters are not pictured in the *Lithography*.

Soon after the *Lithography* was published, Beringer came to realize the hoax, and when the perpetrators then spread rumors accusing *him* of the fraud, he requested a special judicial hearing held in April 1726 for "the saving of his honor." He was exonerated by the testimony of Christian Zänger, one of Beringer's teenage diggers, who confessed his complicity and fingered Roderick, von Eckhart, and the Baron, whom he overheard hatch the plot to ruin Beringer "because he was so arrogant and despised them all." Von Eckhart died a few years later and the Baron escaped the scene unscathed. But Herr Roderick, the ringleader of the plot, absented himself from Würzburg, evidently banished from the Duchy of Franconia.

- Beringer *did* discover Hebraic lettering among his "fossils." But the discovery was made before, not after, publication of his opus; the writing was also Latin and Arabic, not only Hebrew; and the letters spelled out "Jehovah," not Beringer's name.

- Original copies of Beringer's treatise *are* of almost legendary rarity, greatly prized by bibliophiles. A spurious "second" (title-page) version, issued in 1767 evidently at the behest of Beringer's heirs, is not nearly so scarce. But only a few copies of the 1726 first edition have been available for sale in recent decades, exclusively at booksellers' auctions (where their price skyrocketed from $125 in 1961, to $250 in 1971, to more than $1,000 in 1987).

Why Did Beringer Blunder?

It is no surprise that Beringer and his book have long been remembered. It is the story of an overzealous, imperious, and incredibly naive "savant" who was unwittingly duped (by trusted colleagues, contemptuous of his arrogance and privileged status) into accepting carved stones as fossils and thereby making a fool of himself in the eyes of the world. The moral of the saga seems inescapable. Yet Beringer's figured stones bear only the most superficial resemblance to fossils (much less living organisms), and it is hard to imagine that anyone could be fooled by "fossilized" stars, moons, comets, and Hebrew (or Latin or Arabic) letters! Moreover, Beringer assuredly was no bumpkin. He held high positions both in the university and at the Court of Würzburg; he was praised in his lifetime as an "illustrious . . . most inquisitive scholar"; and even in the 1880s, 150 years after the fiasco, he was still acclaimed in the official history of the university as a "virtuoso . . . tireless scholar . . . the most active man of his time." What led this touted academic to ruin?

To fathom Beringer's blunder we first need to understand what was known and what was not known about the nature of fossils. His was a time of turmoil in paleontology, of competing ideas and uncertain answers. There was even the question of whether "lithology" (as the study of rocks, minerals, and fossils was then known) was a pursuit worthy of learned men, a question raised by Beringer himself in chapter 1 of his remarkable tome (and rendered from his stylized, flowery Latin by M. E. Jahn and D. J. Woolf in *The Lying Stones of Dr. Beringer*, a classic published in 1963 by the University of California Press, Berkeley):

> To what purpose, they ask, do we stare fixedly with eye and mind at small stones and figured rocks, at little images of animals or plants, the

rubbish of mountain and stream, found by chance amid the muck and sand of land and sea? To what purpose do we, at the cost of much gold and labor, examine these findings, describe them in vast tomes, commit them to engravings and circulate them about the world, and fill thick volumes with useless arguments about them? What a waste of time and of the labors of gifted men to dissipate their talents by ensnaring them in this sort of game and vain sport! Does this not amount to neglecting the cares of the realm to catch flies . . . the efforts, the genius, and the expenses of learned men gone mad?

Beringer's answer shows a devotion to knowledge motivated by strong religious underpinnings:

The mind of man, [however], was made for higher things, not for the fattening of the belly, nor for the luxury and delights of the body, nor for that most inane occupation of all, the custody of a pile of gold or brass. . . . We do not recognize as scholars of the fine arts those who are the slaves of lucre . . . but rather those noble souls who expend their energies . . . stimulated by the very dignity of learning and of laudable work, in order to experience the pure joy born of knowledge. This is the core of Christian ethics. . . . What more noble purpose for human actions can be conceived than that whereby from the marvelous effects of nature we ascend . . . to the recognition of the power of the Creator?

Cabinets of Curiosities

In Beringer's day, the study of fossils was carried out mostly by physicians, almost always as a hobby when duties and time permitted. Many of these learned dilettantes assembled "cabinets," private collections of natural rarities—rocks, minerals, shells, feathers, animal skins, archaeological oddities. Those with especially noteworthy cabinets often bequeathed them to local municipalities, collections that even today can be ferreted out among the holdings of major museums. One of the earliest cabinets in Europe was that assembled beginning in 1575 by Bernard Palissy (1510?–1590), a Parisian naturalist and talented potter known not only for his splendid crockery (decorated with reptiles, insects, and plants) but also for his outspoken detractors (diluvialists who reviled him for asserting they had "erred clumsily" by claiming fossils as evidence of the Biblical Flood). By the mid-1600s private collections of natural rarities had come to be fairly common. No fewer than forty-one cabinets are mentioned by Maximilian Mission (1650?–1722) in *New Voyage to Italy,* the chronicle of his 1688 travels through Europe.

There is no evidence that Beringer's cabinet—"gathered on nearly all the shores of Europe" and consisting mostly of "oryctics," things dug from the earth—was in any way remarkable prior to May 31, 1725. But on that day his diggers found and delivered to him the first three figured stones, one bearing a "circle, like the sun with its rays," the others wormlike shapes. Within only a few months, hundreds of stones, some having figures on both sides, were added to his collections. Beringer's cabinet had come to be remarkable, indeed! Was Beringer pleased? Thrilled? Ecstatic? You bet! From the Introduction to *Lithographiae Wirceburgensis*: "These wonderful exhibits reveal the most bountiful treasure of stones in all of Germany . . . at last, thanks to my perservering effort . . . discovered and unearthed at no small cost and labor. I doubt that any more heartwarming spectacle can come to the eye of a scholar of natural science." Though Beringer was certain his finds were momentous, he was at the same time deeply perplexed by what to make of them. So, as he reports in the *Lithography*, he did what any scholar would do—he trekked off to the library and looked up relevant works to see whether anyone else had come upon stones like his and a way to explain them. His library research led him to conclude (correctly) that none of the theories fit his finds, and his outstandingly thorough argumentation makes this part of the tome especially valuable, one of only a few surviving sources to tell us how fossils were viewed at that time.

Theories on the Nature of Fossils

Only a few scholars of the time knew as we do now that fossils are the bodies or impressions of organisms preserved naturally in rock. Of these, probably the best known today is Robert Hooke (1635–1703), the first Curator of Experiments to the Royal Society of London and remembered in modern textbooks as a pioneering microscopist. But Hooke's view, far from prevalent in the early 1700s, is largely ignored by Beringer, who deals chiefly with three notions that were much more popular:

1. The theory that fossils are so-called sports of nature—stones formed by an extraordinary "plastic power" present in the Earth.

2. The idea that fossils grew in place from the seeds and eggs (the "Spermatick Principle") of plants and animals lodged in crevices by rain or wind and activated by heat, saline moisture, or other postulated "life-giving force."

3. The concept that fossils are remains of organisms drowned and buried during the Biblical Flood.

Fossils as Sports of Nature

This first theory stems from the writings of Georgius Agricola (1494–1555)—the Latinized name of Georgius Bauer—a German physician and the "Founder of Mineralogy" who viewed fossils as superficially lifelike objects formed in rocks by the action of a "lapidifying juice" (the *succus lapidescens*). In the seventeenth century, Agricola's formulation was updated, and in 1671 it was backed by no less than the house physician to Queen Anne, a respected expert on seashells and fellow of the Royal Society, Dr. Martin Lister (1638–1712): "There may all Manner of Sea-Shells be found promiscuously included in Rocks or Earth, and at good Distances too from the Sea. But . . . I am apt to think, there is no such Matter as petrifying of Shells in this Business: But that these Cockle-like Stones everywhere . . . [were] never any Part of an Animal. . . . There is no such Thing as Shell in these Resemblances of Shells."

Probably the last influential proponent of the nonbiologic origin of fossils was Dr. Robert Plot (1640–1696; fellow and secretary of the Royal Society and editor of its *Philosophical Transactions*), who in an article of 1677 agrees with Lister that some fossils are the work of a "plastick power" (though he opines that others "relate to the Heavenly Bodies" or the "Watery Kingdom").

The Spermatick Principle

The second idea, that fossils grew in place from the "Spermatick Principle" (seeds and eggs) of plants and animals lodged in rock crevices by rain or wind, apparently originated with Andreas Libavius (1540?–1616). Its greatest advocate was Edward Lhwyd (1660–1709), an authority on fossils, antiquities, and Celtic philology who was Plot's successor as keeper (curator) of the Old Ashmolean Museum at Oxford University. Lhwyd's theory is outlined in a long letter written in 1699 to John Ray (1627–1705), an English naturalist whose classification of plants and animals laid the foundation for the system we use today. In Lhwyd's words:

> I have in short imagin'd [that fossil fish] might be . . . owing to fish-spawn, receivd into the chincks . . . of ye earth . . . [and that] the exhalations which are raisd out of the sea and falling down in rains, fogs, &c. do water ye earth to ye depth here requir'd [so that] from the seminium or spawn or marine animals [the rocks become] so far impregnated with . . . [microscopic] animalcula (& also separat or distinct parts of them) as to produce these [fossil] marine bodies, which have so much excited our imagination, and indeed bafl'd our reasoning throughout the whole globe of the earth. I imagind farther that the like origin might be ascrib'd to the mineral [that is, fossil] leavs and branches, seeing we find that they are for the most part the leavs of ferns . . . & such like plants . . . whose seeds may be easily allowd to be washd down by the rain into the depth here requir'd, seeing they are so minute, as not at all to be distinguished by the naked eye.

Lhywd's theory was short lived, last championed by Beringer's contemporary, Carolus Nicolaus Lang (1670–1741), a prominent Swiss physician who, like Lhwyd, believed that fossils originate by germination of the "seeds" of animals deposited in rock cracks by air or water and stimulated by a so-called *aura seminalis*—an "activator" such as subterranean heat or fluids, or, especially, percolating snow water (called on to explain the presence of fossil seashells in mountainous terrains).

Fossils from the Great Flood

The third theory, the Diluvialist hypothesis, held that fossils are remains of organisms buried by the Noachian Flood. Among its most persuasive advocates was John Woodward (1661–1727), "Professor of physick" at Gresham College, Oxford; a fellow of the Royal Society; and a gentleman "extraordinary ingenious . . . who had ye most considerable collection of English fossils." Woodward's views are set forth in his 1695 *Essay toward a Natural History of the Earth.* Though they are diluvialist to the core, they differ from the normal version by arguing that the waters of the Great Flood had the property of "totally dissolving" the Earth's crust, a supplement to diluvialism Woodward thought was necessary to explain the presence of fossils "if not quite down to the Abyss, yet at least to the greatest Depth we ever dig."

Woodward was apparently the first to deduce that the 40 days and nights of "the Deluge commenc'd in the Spring-season: the Water coming forth upon the Earth in the Month which we call May." His ideas influenced many respected scholars (among them the famed discoverer of *Homo diluvii testis*, Johann Jacob Scheuchzer), all of whom recognized fossils as remnants of past life but fit them to theology rather than worldly processes.

Beringer's View

As Beringer struggled to solve the puzzle of his extraordinary finds, he considered less popular notions as well—that the figures on his stones were formed by the influence of the heavens (essentially an astrological explanation), that they were impressed on them by a "plastic power" of light, or that they were the handiwork of prehistoric pagan tribes. Though the last came closest to solving the riddle, Beringer dismissed it (since pagans could not know the name Jehova) as well as all other explanations as not fitting the evidence at hand: "I attempt to demonstrate . . . that the nature of our stones is so unusual that its very novelty eludes the [above-noted] opinions, however well established by documentation and searching experimentation [they may appear to be]." Until the fraud finally came to public notice, he remained firm that the stones were "works of Nature." Since he was at a loss to explain how nature made them, he ended his treatise by begging the question, arguing that judgment of his "earthly treasures" should best be left to "wise men . . . to the men of letters, to the scholars and patrons of the finer and more profound disciplines" who in time would render a final verdict.

Unearthing a Rosetta Stone

In 1725 Beringer thought he had discovered a Rosetta Stone, though he couldn't make out its meaning. In the very same year, Scheuchzer was *certain* he had unearthed a Rosetta Stone, one linking the biblical past to the present. Beringer may have been naive to be so fooled, but he was scholar enough not to claim more for his find than he could fathom. Scheuchzer shoehorned his to fit accepted dogma, an error that held on for nearly 100 years. In the 20/20 hindsight of two-and-a-half centuries, both can be seen to have missed the mark.

Yet who are we to smuggly sit in judgment? It is of course true that knowledge has grown immeasurably since the days of Beringer and Scheuchzer, but it is well to remember that the human side of science has largely stayed the same. Like these old-time scientists, we carry out our daily work saddled with ideas that we have been taught are true and struggle just as they did to make lasting contributions. And we are probably no more immune than they from the heady pull of a major breakthrough or even the urge to shoehorn finds to fit prevailing views. Moreover, though it is now harder to be fooled since so much more is known, it's a sure bet that some of what passes as "known" today will eventually turn to dust. Still, we can take heart, for Beringer and Scheuchzer show that even when human foibles sidetrack the search for knowledge, the path will be regained. We can only wonder what the scholars of the year 2250 will think of us. But one thing is certain—by then, *our* most glaring blunders will have long since been cast aside.

12

The Hunt for Life on Mars

You must not fool yourself—and you are the easiest person [for you] *to fool.*
Richard Feynman, Cal Tech, Commencement Address, 1974

Hints of Ancient Martian Life?

The news was trumpeted in banner headlines: "PAST LIFE FOUND ON MARS!" Asked by NASA to give the public a first-blush scientific appraisal of this extraordinary claim, I was on hand at the Washington, D.C., news conference that announced the find in August 1996.

But my involvement dates from more than a year earlier, when in January 1995, at the request of NASA administrators, I journeyed to the Johnson Spacecraft Center (JSC) in Houston, Texas. My mission was to render a verdict on what geologists there believed might be microfossils in a chunk of meteorite thought to have come from Mars. Designated the judge of their possibly breakthrough discovery, I was sworn to secrecy by the JSC scientists lest their find hit the newspapers before they had the facts. What caused the fuss were tiny orange-colored, pancake-shaped globules of carbonate mineral, 2 to 200 μm across and ringed by thin black and white rinds. Flushed with excitement, the researchers claimed that never before had ringed disks like these been seen in a meteorite. Since this one was said to have come from Mars—which in the distant past may have harbored life—and since the objects were made of the same mineral and some were about the size as shells of a particular type of protozoan (foraminiferans, "forams" for short), they thought they might have chanced on a mélange of martian fossils. Because of my work on the most ancient fossils known on Earth (chapter 3), I had been brought in to shore up the paleontologic guess of geologists schooled in rocks and minerals, but not biology.

Their guess was wrong. The globules certainly were not remnants of protozoans. A number of the objects were simple foramlike disks but many others merged one into another in a totally nonbiologic way. Their overall size range also did not fit biology, and they lacked any of the telltale features—pores, tubules, wall layers, spines, chambers, internal structures—that earmark tiny protozoan shells. Moreover, the "lifelike" traits they did possess (carbonate composition, discoidal shape, ringed rims) could be explained by ordinary inorganic processes.

Carbonate minerals are, of course, laid down by life—by protozoans, clams, snails, corals, even certain seaweeds. But carbonates also form by purely inorganic means and are known from many meteorites, not just the one containing the putative fossils, where their nonbiologic genesis is beyond question. The discoidal shape of the carbonate globules didn't seem to require biology, either. Formed when mineral-bearing solutions percolated through a thin crack in the rock, the pancakes are flat on both top and bottom because the solidifying carbonate ran out of space above and below. And they are more or less circular and merge one into another: as minerals drop out of solution, they crystallize around all sides of a grain first formed (the "center of nucleation"), in this case making a disk or, if the grains are closely packed, discs that merge. Their rimming rings, I thought, came from the same process. When the makeup of a crystallizing solution changes, so do the minerals laid down. The thinly layered black and white rims showed that chemical conditions changed as the pancakes formed, not that the disks were shells of protozoans.

I raised these points with the JSC scientists. They seemed to agree. The matter, I thought, was closed. But I urged them to continue the hunt. I believed then, as I do now, that the search for hints of life in martian meteorites is a promising way to attack a truly fascinating question. It *is* important to know whether life once existed (or still does) on Mars. (Still, I was taken aback when more than a year later, at the August 1996 news conference, the same little pancakes were again proffered as evidence of martian life, this time of bacteria rather than "protozoans." Evidently the scientists' minds were set—the facts hadn't changed, only the meaning attached to them.)

NASA Stages a Press Conference

Several weeks before the August news conference, I received a phone call from NASA headquarters informing me that the JSC scientists I had visited earlier had completed studies of a meteorite they claimed held evidence of ancient life on Mars. A technical article reporting their results was soon to appear in *Science*, a highly respected journal reserved for the hottest of hot discoveries. NASA felt obliged to inform the public and planned to do so at a prepublication press conference. But because some at headquarters thought the evidence "a bit iffy," they wanted an outside expert to publicly evaluate the findings when they were announced to the world. Would I, please, perform this task?

I was reluctant. I had plenty on my plate already and feared this was one more in a string of spurious claims for "life in meteorites" that dates to the early 1960s. Still, I hadn't read the article, hadn't seen the evidence. And the scientists making the claim were colleagues. I agreed to "think about it."

A copy of the soon-to-be-published report arrived the next day. I studied it. Carefully. Three times. I was not impressed. Though some of the report was backed by solid scientific data, support for other parts was wanting. Crucial questions had not been asked. Articles published earlier and critically relevant to the authors' contentions had been ignored. More plausible and alternative ways to explain the findings were given short shrift. The manuscript's concluding claim of "evidence for primitive life on early Mars" seemed overblown, ill-conceived.

I called NASA and, quoting Carl Sagan's catch phrase that "extraordinary claims require extraordinary evidence," suggested that for this claim the evidence was not even close. I offered names of three other scientists to serve in my stead. A few days later NASA called back and informed me that the agency's director, Dan Goldin, had personally pegged me for the job—partly, I gather, because he's a Sagan fan (and was said to have been pleased by the quote), but I think mostly because he knows it's in NASA's best interest to get the story straight. Any claim for life on Mars—whether of organisms small or large, past or present—is bound to stir controversy. This one would be no exception. The "iffy" evidence was certain to raise eyebrows, and since NASA's budget hearings were looming, even the timing of the announcement might be regarded as suspicious. My guess is that Mr. Goldin—a truly able administrator and brilliant politician (appointed by Republican Bush, a star of Democrat Clinton's team)—figured a preemptive strike was in order. To protect NASA's reputation and at the same time stifle the easily predicted army of naysayers, he decided to assign a hard-nosed outsider to evaluate the claim. Who better than one calling for it to be backed by "extraordinary evidence"?

Before I took the call relaying Mr. Goldin's personal request, I thought I was in the clear. This was a task I did not want to do. But Goldin is the NASA boss—the "faster, cheaper, better" guy, an appointee of two presidents. Who was I to turn him down? I agreed.

Prelude to the Feeding Frenzy

The news conference wasn't scheduled for another two-and-a-half weeks. I tried to put it out of my mind, but it began to gnaw at me. My skepticism was bound to raise some hackles. I spent the next weekend listing my arguments on vu-graphs (see-through charts like NASA often uses), and the following Tuesday morning I faxed copies to Houston. It was only fair to warn the JSC group what I planned to say, and I also wanted to make certain I had not misunderstood the technical details of their article—I was sure they'd straighten me out.

My hope for dialogue came to naught. Neither they nor I had time. About an hour and a half after I sent the fax, I received yet another call from Washington: "Bill, get on the 1:30 afternoon flight. There's been a news leak—the press conference has been moved up."

I arrived at Dulles Airport late that night and at NASA headquarters the next morning. There I was squired to a basement room, where I found the JSC team seated side by side, rehearsing their lines one after the other. They were prepared. Thoroughly. They even had high-tech cartoon videos to tell the story of the flight of the meteorite from Mars through space, and how they had made their finds. And though the room lacked a VCR to show the videos, the group didn't skip a beat. When they came to the video part of their run-through, one of the team said: "My video talkover lasts 2 minutes, 47 seconds." The one next to him laughed: "Mine's only 2 minutes, 19." (VCR-blind, the first member delivered his rendition practically the same as he gave it later to the reporters upstairs. The 2-minute, 19-second version changed not at all. These folks were pros!)

Finally, it was my turn in the practice session. They had videos. I had vu-graphs. They had practiced. I had not. They were NASA. I, an outsider. Still, I gave my spiel. By that time, there was a pride of NASAites overlooking our run-through, Dan Goldin among them. I finished. Silence. Utter stillness. Then a woman from the headquarters staff rose and berated the troops: "Schopf has just demolished you. Can't you guys be more *positive?*" (I don't know who this person was—was never introduced, never caught her name—but you can see her on the CNN tape of the press conference introducing Administrator Goldin.) The JSC crew was in a quandary. Like I, they knew their story was circumstantial. There was no "smoking gun." But it was important for them to look good, to please the boss. The pressure was great. They seemed torn.

At the practice session, I had tried to be reasonable, even gentle. I did, too, at the later news conference, a performance for which I've been much praised—but I was also chastised (by no less than a Nobel laureate) for being too soft. Still, it seems to me that the Mars Meteorite Research Team (as they were now calling themselves, bolstered by input from scientists at McGill, Georgia, and Stanford Universities), tackled a difficult interdisciplinary problem. An instant answer, pro or con, was not in the cards.

Breaking the News to the World

Not only had I not practiced for the news conference, I had not been warned what to expect. Maybe no one knew. The only thing I had to go on were memories of the late 1960s when five other scientists and I (officially, the Lunar Sample Preliminary Examination Team) were tasked to do the first studies of Moon samples gathered on the Apollo 11 and 12 missions. While the Apollo crews rested in quarantine in another part of the building, we sorted, studied, and described the rocks. To test whether they harbored virulent Moon germs (which we playfully dubbed "Gorgo"), we even monitored the effects of lunar dust fed to Japanese quail, germ-free mice, and various plants (some of which grew better on moon dust than on Earth soil). Interactions with the media were friendly; interviews were one-on-one or at most with a few pool reporters from magazines, newspapers, radio, TV.

The Mars news conference could not have been more different. Instead of only a few reporters, there were five-hundred. Instead of note pads, there were scores of video cameras. There was so much electronic gear in the auditorium that the sound system overloaded and the conference had to be delayed to take care of high-pitched feedback whining through the hall. On the stage I was seated alongside the chief of the Stanford group that identified organic compounds in the meteorite. Just before the conference was to begin, he waved to a friend among the gaggle of journalists. Within only a few seconds he, and I next to him, were besieged by a churning sea of microphone-thrusting reporters, each fixed on elbowing to the front of the pack. A media feeding frenzy!

Things quieted down and we waited for another 20 minutes as coverage switched to the South Lawn of the White House, where President Clinton read a carefully crafted statement on the signficance of the find about to be unveiled. For the next two and a half hours CNN carried our press conference to the world. Mr. Goldin led off, followed by the well-choreographed presentations of the Research Team. My remarks came last, followed by a lengthy session of questions from the Washington press corps and journalists gathered at NASA installations across the country, and answers from those of us at the dais.

The team's presentations were measured, sensible, their arguments plausible. By the time they finished, virtually the entire throng of journalists on hand seemed willing to believe. Introduced as the designated "skeptic" to "begin the debate," I had no doubt my words would prove unwelcome. I was like Daniel in the lion's den. But the evidence was (and still is) inconclusive, and it fell on me to point that out. Some claim the glass is half full; to others it's half-empty. But no one who knows the facts would claim the glass is overflowing—not then, not now. Not even the Mars Meteorite Research Team.

Meteorites from Mars

Mars as an Abode for Life

In some ways Mars is like a smaller version of our planet, a sunlit rocky body of half the size and one-third the gravity, and with a day only 37 minutes longer and seasonal swings (in a 669-day year) like Earth's. But in other ways, Mars differs. It has ancient rusty plains pockmarked like the Moon, and the longest (4,000 km) and deepest (nearly 10 km) canyon—Valles Marineris—known anywhere. It boasts a volcano some 40 kilometers high (Olympus Mons), the largest ever seen, as well as sinuous channels carved by rivers now long dry. It never rains on Mars, and its mostly (95%) carbon dioxide atmosphere is so thin, one-tenth the pressure of Earth's, that without the cocoon of a protecting spacesuit one's eyes would pop out. Practically the entire planet is a frigid wasteland. Even near the equator, for instance at Carl Sagan Station where NASA's Pathfinder landed in July 1997, temperatures range during the martian summer from that of a freezing winter day in Fargo, North Dakota, to the coldest on Earth.

Yet not always was Mars so cold and dry. Though the planet probably was never close to tropical, its youthful climate was much more hospitable to life like ours—a fact key to the past-life-on-Mars story, since the martian meteorite dates from early in the planet's history when rivers flowed, the atmosphere was thicker, and life may have gained a foothold. A second key is that the story centers on minute forms of life, on bacteria, the simple single-celled microbes that play a far larger role in the evolutionary Tree of Life and are much more resilient than previously thought. They exist on Earth in a striking range of settings—scalding deep-sea vents, sulfurous acid springs, cracks and crevices in rocks deep in the crust, on and within ice sheets and permanently frozen Arctic tundra, even in mineral-encrusted fissures in the rocks of Marslike ice-cold deserts. If bacteria can survive, even thrive, in such settings here, why not also on Mars?

ALH84001

The claim for ancient life on Mars comes from a potato-sized, 1.9-kilogram meteorite, ALH84001—named for where and when it was found (Allan Hills ice field, Antarctica, in 1984) together with its

Figure 12.1 The Antarctic meteorite ALH84001; the linear striations are saw marks made when the meteorite was cut for sampling; they are not layers in the rock. (Courtesy of NASA.)

assigned sample number (001)—that was plucked out of the ice by a young geologist named Roberta Score (a graduate of my UCLA department) during an annual expedition of the National Science Foundation's Antarctic Search for Meteorites program.

Dates measured by radioactive isotopes show that the rock was formed at nearly the same time as Mars was born, 4.5 billion years ago, and the makeup of the meteorite places its genesis a few kilometers deep within the congealing crust. Though sketchy, its subsequent history can be pieced together. Like early Earth, ancient Mars was bombarded by rocky chunks swept from orbit as it circled the Sun. According to the scenario favored by the research team, these impacts cracked and fractured ALH84001, and since this was early in the planet's history—3.6 billion years ago, when Mars was warmer and wetter—groundwater seeped through the fissures and filled them with carbonate mineral. But the age of the fracture fillings is open to question, by some evidence dating from only 1.3 rather than 3.6 billion years. An alternative version has the mineral emplaced much later, when the impact that careened ALH84001 off the martian surface infused the veins with hot carbonate-charged fluids.

About 16 million years ago an asteroid struck Mars with terrific force, gouging a huge crater and ejecting pieces of the planet's surface with enough power to escape its gravitational pull. One of those pieces was ALH84001. It hurdled through space for millions of years until it felt Earth's tug and fell to Antarctica 13,000 years ago.

Rocks like ALH84001 are as rare as hen's teeth. Though thousands of meteorites are known to science, it is one of only twelve identified as martian. A chemical signature (a mix of the isotopes of oxygen, $^{16}O/^{17}O/^{18}O$) in minerals of these twelve shows they are not Earth rocks and not from the Moon, and because they share the same chemistry all are thought to come from the same source. Like the others, ALH84001 is an igneous, once-molten rock, so it and the others must have formed on a body large enough to have partly melted. Planet-sized masses fill the bill, and with Earth and Moon ruled out, only Mars, Venus, and Mercury are left. The link chaining the group to Mars is provided by one of ALH84001's siblings (meteorite EETA-79001), which contains tiny pockets filled with gases that match those measured in Mars's atmosphere by NASA's 1976 Viking landers. The gas mix is distinctly Marslike and differs from any known elsewhere.

Rocks Trickle in from Mars

Most meteorites are debris left over from when the solar system formed. But about two dozen have been identified as chunks dislodged from planetary neighbors, half from Mars, half from the Moon. Six of the twelve Mars meteorites were discovered in Antarctic ice fields, so of the eight-thousand meteorites recovered from Antarctica roughly one of every one-thousand is a piece of Mars.

Though ordinary travel times from Mars to Earth are millions of years, under some conditions they can be very much less. Using computers, scientists at Cornell University simulated the histories of more than two thousand objects careened off Mars's surface and found that a small fraction could have arrived in no time flat. According to their calculations, "fast transfers (taking less than a year) from Mars to Earth must have occurred numerous times during the Earth's past. . . . If martian microorganisms can survive a year in space, many may have already arrived." The Cornell group focused on *live* organisms, whereas the Mars Meteorite Research Team's evidence is of life long dead. If live microbes could get here, their fossils might too!

Search for the Smoking Gun

The claim for ancient life on Mars is backed by three kinds of evidence, all found in the carbonate-filled fractures of ALH84001:

1. Tiny pancake-shaped globules—orange-colored carbonate disks and their dark (iron sulfide and oxide) rims—made of minerals that on Earth can be formed by bacteria.

2. Organic molecules like those produced by breakdown and geologic aging of fossilized organic matter.

3. Minute jelly-bean-shaped and threadlike objects that resemble fossil microbes.

At the NASA press conference, Administrator Goldin proclaimed the findings "compelling." In one sense they certainly are, for like clues in a good detective story they are captivating, even gripping. But the findings are far from "irresistible, overwhelming," as the term is also used, and of the three lines of evidence only one, the possible fossils, is a potential smoking gun.

Martian Minerals

Consider first the martian minerals. On Earth, bacteria sometimes play a role in forming carbonate, sulfide, and oxide minerals like those in ALH84001. Yet the same minerals are common products of geology, made by wholly inorganic means, and all are present in other meteorites where their nonbiologic source is beyond dispute. The minerals hold clues to the history of the Mars rock but are themselves not firm evidence of life.

Nothing about the carbonate pancakes pegs them as products of biology. Resurrected from their earlier claimed identity as possible "martian protozoans," their link to bacteria is equally unproven. The sulfides also lack a biologic signature, and the mix of sulfur isotopes they contain would on Earth tag them as nonbiologic rather than bacterial. And though the iron oxides (minute crystals of magnetite, Fe_3O_4) have been dubbed "magnetofossils" in the popular press and in some scientific articles, they are fossils in name only. Crystals like those in ALH84001 are present in certain strains of microbes that use them as tiny compasses if they are the right size (40 to 120 nanometers, or nm), have their magnetic poles pointing in the same direction, and are linked in long ensheathed chains that boost the magnetic signal. The iron oxides in the martian meteorite come in other sizes as well, are randomly oriented, are never bound in chainlike aggregates, and cannot be told apart from ordinary grains of magnetite formed by geology.

A piece of the puzzle not yet in place is the temperature at which the minerals formed, a crucial detail because life's chemistry breaks down if temperatures are too hot. The current world's record of 113°C is held by the superheat-loving (thermophilic) microbe *Pyrolobus* that was isolated from deep-sea fumaroles, and key molecules of life begin to fall apart above about 130° C. The Meteorite Research Team argues that since the chemistry of the vein-filling carbonate shows it formed at temperatures low enough (less than 80° C) for life to exist, the carbonate could have been laid down by bacteria. But other workers using different indicators arrive at much higher estimates—150°, 250°, 300°, even more than 650° C—temperatures too hot for life. A high temperature would rule life out whereas a low one would show only that microbes could have existed, not prove they did.

New results can be expected from continuing studies of the mineralogy of ALH84001 and other martian meteorites. But unless the minerals are somehow shown to be definitely biological, they cannot answer the question of life on Mars.

Organic Molecules Possibly from Mars

One of the most intriguing findings of the Meteorite Research Team is the identification in ALH84001 of organic compounds composed of carbon and hydrogen known as polycyclic aromatic hydrocarbons (PAHs). Though this discovery points in a promising direction for future exploration of the Red Planet, the presence of PAHs, like that of the minerals, is not proof of martian life.

As we saw in chapter 4, organic compounds are not always a signpost of life. The organic molecules of carbon-rich meteorites (carbonaceous chondrites), interstellar dust clouds, or those made in Miller-type syntheses, for instance, are formed by nonbiologic chemistry. But though PAHs are also known from carbonaceous chondrites and surfaces of dust grains and graphite particles that cruise through interstellar space (where their nonbiologic origin is not in question), those in Earth rocks come from once-living matter, formed during its geologic breakdown and aging. So while PAHs are not biochemicals, never themselves made by life, they are plentiful in plant fossils, coal, and the small carbon-rich particles of kerogen that give black shales their color. Generated by burning of fossil fuels, they are common also in automobile exhaust and factory smoke. They are even present in the vapors rising from a grilled steak!

If on Earth PAHs come from the breakdown of biologic substances, why not also in ALH84001? The possibility exists but remains unproved because PAHs are made easily by chemical reactions in the total *absence* of life, and there is no way to tell whether they come from life or not (except possibly for their carbon-isotopic makeup, yet unknown for the PAHs in the martian meteorite). Like the PAHs in other meteorites, those in ALH84001 may be entirely nonbiologic.

There is yet another problem. Ever since the NASA press conference there has been debate whether the PAHs found in ALH84001 actually belong to the martian rock. Although there are good reasons to believe they were in the rock when it crashed to Earth, it's not altogether certain: because PAHs from atmospheric pollution and probably also from Antarctic coals are present in the snow and ice at Allan Hills, and those in the meteorite are on surfaces of cracks where they may have been deposited from seeping meltwater. But even if the PAHs are actually martian, they can never show there was life on Mars unless the PAHs are proven to come from biology.

Still, if they are truly from Mars, the PAHs could pave the way to important findings. NASA plans to hunt for life in Mars samples slated to be returned to Earth in 2008. Only hard-line enthusiasts think the dust and stones will harbor anything alive. Unlike the times of its more clement past, Mars now is an awful place to live: its surface is drenched in lethal UV rays and it is so dry and frigid that there is no water for life to use. So NASA has pinned its hopes on detecting stromatolites, and within them microscopic fossils, a needle-in-a-haystack hunt. The first task will be to find the haystack, and organic compounds like PAHs may provide the means. On Earth, organic matter and life go hand in hand. The same should hold for Mars if life ever existed there. NASA's best bet is to ferret out and bring back rocks rich in coaly carbon and search them for tiny fossils.

Fossil Microbes on Mars?

Ancient life on Mars. The minerals can't prove it. The PAHs can't, either. The "fossils" could—but they don't, and there are good reasons to question whether they are in any way related to life.

"A picture is worth a thousand words." Shown in newspapers, magazines, and on TV around the world, "Mars fossils" have captured the public's fancy. Their lifelike shapes are palpable, far easier to understand than arcane chemistry. But unlike the press, the Mars Meteorite Research Team has handled the supposed fossils with kid gloves. Their seven-page technical article includes only four sentences on the objects which suggest, almost as an afterthought, that as "features resembling terrestrial microorganisms . . . or microfossils [they are] compatible with the existence of past life on Mars." The objects they illustrate are exceedingly small, 20 to 30 nm across, and though their jelly-bean-like form *does* resemble rod-shaped bacteria, they are much too minute for the comparison to hold. More than a million times smaller than a run-of-the-mill bacterial cell, they are actually more like ribosomes, 20-nm protein-making bodies present in cells in prodigious numbers—nearly 20,000 in a typical bacterium, more than 100,000 in a human cell. The "Mars fossils" are the size of minute particles *within* bacteria, not bacteria themselves!

Other than shape, size is the only hard fact revealed about the fossil-like objects, potentially telling because there is a limit to how small cells can be. The tiniest microbes on Earth, bacteria of the genus *Mycoplasma* that live as parasites in cells of other organisms, usually mammals, show the limits of life. The most minute are about 0.1 μm (100 nm), contain only a fraction of the genes of a typical

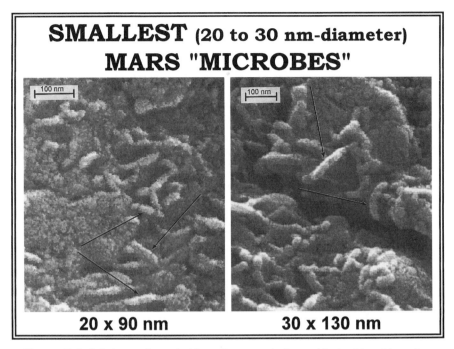

Figure 12.2 Jelly-bean-shaped objects (at arrows) pictured in the Mars Meteorite Research Team's article.

bacterium, and are encapsulated by a thin membrane rather than a sturdy cell wall. Lacking much of the machinery present in free-living microbes, they are able to function only because they are bathed in the nutrient-rich cytoplasm of the host cells they inhabit.

If the "Mars fossils" actually were once alive they must have been composed of cell-like compartments that cordoned off their living chemistry from the surroundings. But because their 20 to 30 nm breadth is thinner even than the simplest bacterial cell wall, they would have been like mycoplasmas, bounded by a membranous structure rather than a thick-walled casing. Judging from biological membranes that take up 6 to 10 nm, the living cavity would have been minuscule. No information has appeared about what encases the "fossils" (or even whether they have true cells), so the space available to house living juices can only be estimated. But it is certain to be much less than that of the tiniest mycoplasma, evidently by about two-thousand times. In other words, for the "Mars microbes" to grow

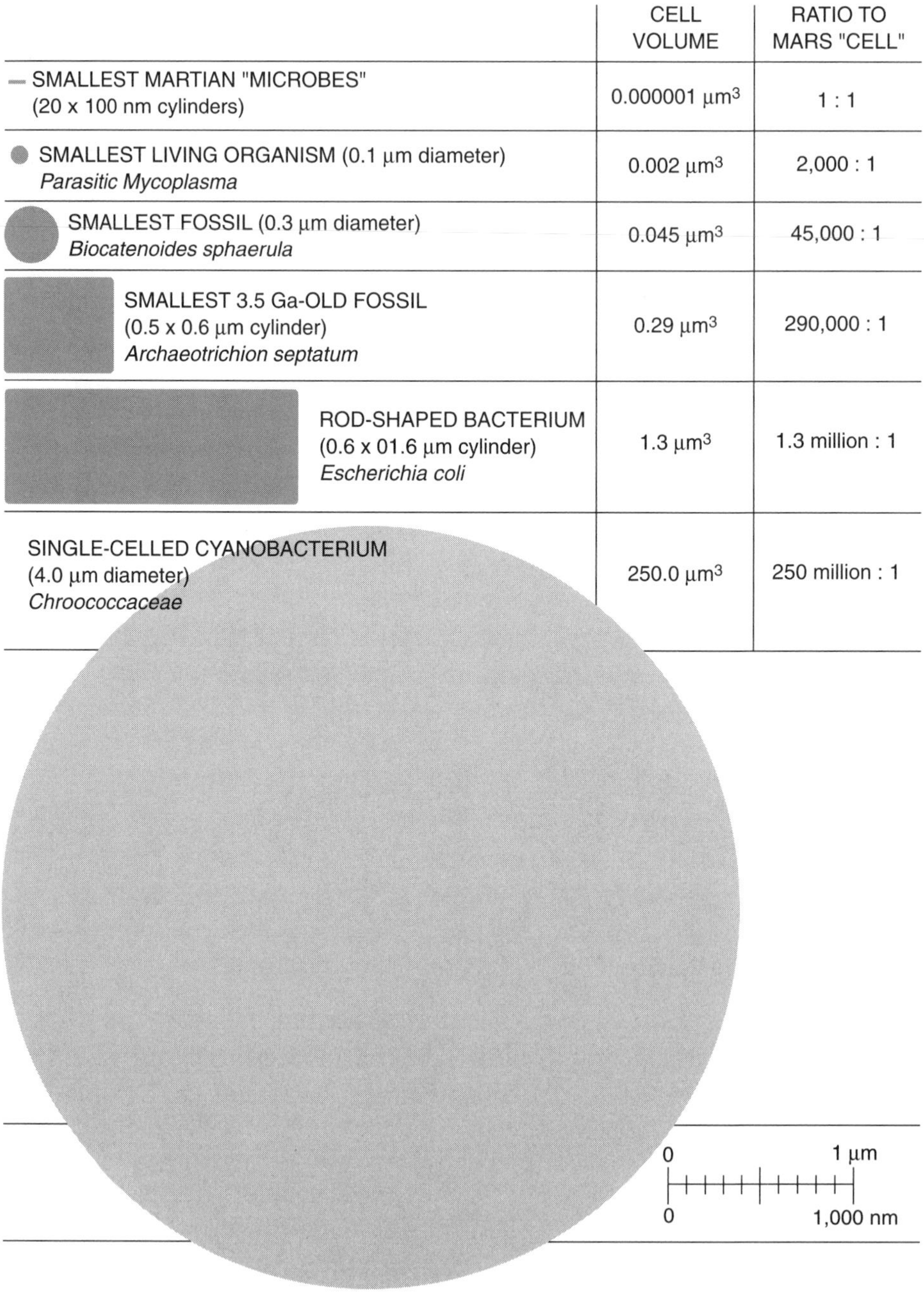

	CELL VOLUME	RATIO TO MARS "CELL"
— SMALLEST MARTIAN "MICROBES" (20 x 100 nm cylinders)	0.000001 µm³	1 : 1
● SMALLEST LIVING ORGANISM (0.1 µm diameter) *Parasitic Mycoplasma*	0.002 µm³	2,000 : 1
SMALLEST FOSSIL (0.3 µm diameter) *Biocatenoides sphaerula*	0.045 µm³	45,000 : 1
SMALLEST 3.5 Ga-OLD FOSSIL (0.5 x 0.6 µm cylinder) *Archaeotrichion septatum*	0.29 µm³	290,000 : 1
ROD-SHAPED BACTERIUM (0.6 x 01.6 µm cylinder) *Escherichia coli*	1.3 µm³	1.3 million : 1
SINGLE-CELLED CYANOBACTERIUM (4.0 µm diameter) *Chroococcaceae*	250.0 µm³	250 million : 1

0   1 µm
├┼┼┼┼┼┼┼┼┤
0   1,000 nm

Figure 12.3 Comparison of the sizes of living and fossil organisms with the Mars "fossils" reported by the Meteorite Research Team.

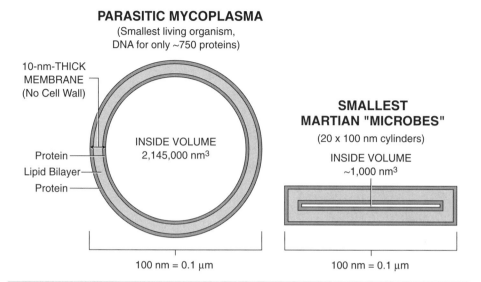

PARASITIC MYCOPLASMA
(Smallest living organism,
DNA for only ~750 proteins)

10-nm-THICK
MEMBRANE
(No Cell Wall)

**SMALLEST
MARTIAN "MICROBES"**
(20 x 100 nm cylinders)

INSIDE VOLUME
2,145,000 nm³

INSIDE VOLUME
~1,000 nm³

Protein
Lipid Bilayer
Protein

100 nm = 0.1 μm

100 nm = 0.1 μm

*MARTIAN CELL VOLUME MORE THAN 2,000 TIMES
SMALLER THAN SMALLEST LIVING ORGANISM*

Figure 12.4 Comparison of the smallest known cellular organism (a parasitic my-coplasma) with the putative fossils reported from the Mars meteorite.

and reproduce even like rudimentary mycoplasmas, each of their bio-chemicals would have to do the work of some two thousand earthly counterpart molecules.

Something is amiss. How could such tiny microbes live? They are said to be rock dwellers, not parasites, not bathed in life-supporting nutrients. Because they are claimed to be "primitive," it is difficult to fathom how their biochemistry could be so efficient (especially in comparison with a mycoplasma's, which is actually a highly evolved version of chemistry handed down from a larger, originally free-living ancestor). And if the "fossils" are actually billions of years old and too tiny to have cell walls, why aren't they weathered, flattened, crushed, wrinkled, or shredded like ancient fossils on Earth?

At the NASA news conference, pictures were unveiled of other "fossils" not mentioned in the published report. Two types were shown: curved simple cylinders, and a stringlike specimen cracked into segments. Though these objects are similar in general shape to earthly

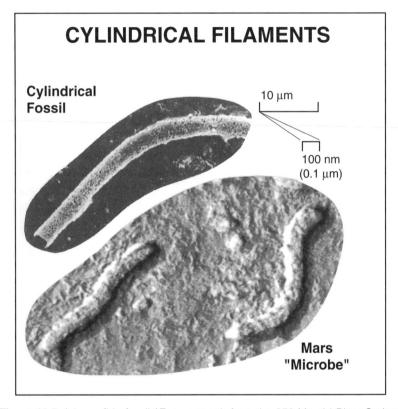

CYLINDRICAL FILAMENTS

Cylindrical Fossil

10 μm

100 nm (0.1 μm)

Mars "Microbe"

Figure 12.5 A bona fide fossil (*Eomycetopsis* from the 850-Ma-old Bitter Springs Formation of central Australia; scale = 10 μm) compared with the cylindrical "Mars microbes" unveiled at the August 1996 press conference (scale = 100 nm = 0.1 μm).

microbial fossils, their volumes are hundreds of thousands of times smaller. Hardly anything is known about either these or the jelly-bean-shaped structures. For example, it is unknown whether they are composed of organic matter rather than mineral. Or whether they are hollow and cellular rather than solid and crystalline. Or why they are exposed in bas-relief on fractured surfaces rather than embedded in rock as Earth fossils are. Their overall size range, variability, abundance, and distribution in ALH84001 has yet to be figured out. In fact, it has not even been shown that they are unquestionably part of the meteorite rather than contaminants or stringy substances splashed on the rock fragment when it was examined.

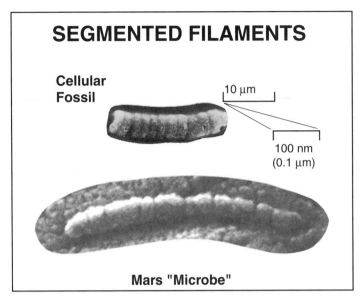

Figure 12.6 A cellular fossil cyanobacterium (from the 850-Ma-old Bitter Springs Formation of central Australia; scale = 10 μm) compared with the single known segmented (cracked) martian stringlike specimen (scale = 100 nm = 0.1 μm).

According to a report in the *Washington Post*, the segmented string-like object, now world famous as the icon of martian life, is the only such specimen ever found—and since it has never been seen again and its exact whereabouts are unknown, no one can check it out further. Another report, from a non-NASA scientist who examined the meteorite firsthand, has it that "the putative microfossils are nothing more than narrow ledges of mineral protruding from the underlying rock, that . . . masquerade as fossil bacteria." Even the Mars Meteorite Research Team seems to have backed off from its earlier claims, in their words warning that "the morphology of the possible fossil forms . . . is certainly not definitive, and more data are needed."

Could Bizarre Mars Fossils Be Identified?

Because microbes can live almost anywhere and are the most ancient forms of life on Earth, it seems sensible to search Mars rocks for bacteriumlike fossils and signs of their living processes. But what if martian "bacteria" were not at all like those on Earth? Could fossil cells truly bizarre in earthly terms be identified as remains of life?

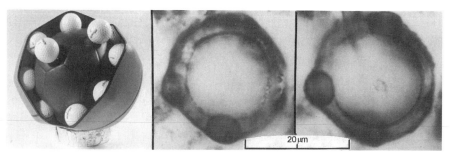

Figure 12.7 A model (*left*) of the bizarre 2,100-Ma-old microfossil *Eosphaera* (*center* and *right;* shown also in plate 1).

This kind of question is not new to paleobiologists, especially those hunting life's remnants in Precambrian rocks. The answer is yes, even for tiny organisms long extinct that bear no obvious relation to life today. Two examples illustrate the point.

Imagine a minute microbe having the form of a soccer ball covered by a layer of scattered golf balls encapsulated by a basketball. Nothing so bizarre exists today. But cells like this floated in shallow seas more than 2 billion years ago. Named *Eosphaera* ("dawn sphere") and known to be planktonic because of its spread in the fossil-bearing rock (the Gunflint Formation of southern Canada), we can only guess that the soccer ball core is a central cell, the golf balls reproductive bodies, the basketball a protective shroud (see plate 1). But we know for certain that *Eosphaera* was once alive—it is made of organic matter (now coaly), has cells and wall layers, is known from many specimens (some complete, others decayed, distorted, torn, flattened), has a biological size range, is part of a complex biological community, was fossilized by processes well understood.

Forms even more other-worldly can be envisioned—for example, chocolate-covered peanuts connected by slender stems to miniature umbrellas. Bizarre indeed! But organisms of this form, too, have been found in the Gunflint rocks and named *Kakabekia umbellata* (for the Kakabeka Waterfall, where the first examples were discovered, and its umbrellalike crown). How *Kakabekia* fits in the Tree of Life is completely unknown, but enough specimens have been found to guess its life cycle (the crown expands from parasol to large umbrella as peanutlike spores are spawned to reproduce the stock).

Figure 12.8 A model of the bizarre 2,100-Ma-old microfossil *Kakabekia*.

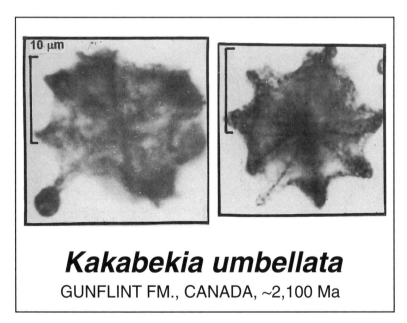

Figure 12.9 Specimens of *Kakabekia umbellata*, a Precambrian microorganism not obviously related to microbes living today.

Life varies over time, from place to place, no doubt from planet to planet. But if the right questions are asked and enough data are amassed, even fossils strange to us can be identified as remnants of life long past.

Lessons from the Hunt

Headlines Win

Perhaps the most obvious lesson learned from this latest chapter in the search for life on Mars is one all too familiar: initially, at least, headlines and soundbites win while facts and reason lose. Most Americans (more than 60% by one poll) agree that "NASA has proved primitive life was present on Mars."

In the face of iffy evidence and a multitude of unanswered questions, why do so many take this view? Some simply want to believe, others are impressed by NASA's track record and think its backing to be foolproof. Even among scientists, few are expert in the distantly related fields needed to assess the report. Yet hopes certainly have been stirred. If the claim collapses after being sold to the public as "good science," it will give a black eye not only to NASA but to science as well.

The Humanness of Scientists

A second lesson is that scientists are no more immune from workplace pressures than anyone else, which can be illustrated by two different readings of the recent history.

By one, NASA geologists chanced on a suite of minerals they thought to be possibly biologic in an ancient martian meteorite. Spurred by these hints they researched meticulously for two and a half years and added supporting evidence from the Stanford PAHs group and specialists at McGill and Georgia universities. Judicious scientists, they released the findings only after their soon-to-be-published manuscript passed peer review and then only at the behest of NASA headquarters, which felt dutybound to inform the public. The published account was meant to be a preliminary report, not the final word, and the claim was of evidence "compatible" with past life on Mars, not that its presence had been proved. But "compatible . . . possible . . .

perhaps . . . maybe" make mushy soundbites and don't sell newspapers. NASA's Research Team was done in by an overzealous press corps.

An alternate scenario has it that the researchers came to be so caught up in the find that normal caution was cast aside. Evidence at odds with the hoped-for outcome was marginalized (such as that of life-searing high temperature and the unbelievably small size of the supposed fossils) or even shoehorned to fit the story (such as the "protozoans" recast as bacterial detritus). To seal the case, eye-catching pictures of fossil-like objects were unveiled to the public without peer review or the backing of solid studies. The announcement of still-preliminary results was premature, but with NASA's congressional budget hearings on the horizon the scientists acquiesced to higher-ups who wanted NASA in the headlines. President Clinton's introductory remarks and the press conference itself were part of an elaborate PR blitz that began a week earlier than scheduled, not so much because the news had leaked (as it surely had), but to avoid being upstaged by presidential candidate Bob Dole's impending announcement of a running mate.

Parts of each version may be right.

Science Is Not a Guessing Game

A third, and probably the most important take-home lesson is that science is self-correcting. There are fine lines between what is known, guessed, and hoped for, and because science is done by real people these lines are sometimes crossed. But science is not a guessing game. The goal is to know. "Possibly . . . perhaps . . . maybe" are not firm answers, and feel-good solutions do not count. Life either once existed on Mars or it didn't. ALH84001 either holds telling evidence or it doesn't. Eventually, hard facts will sort it out.

It is right to demand that extraordinary claims be backed by extraordinary evidence. But in the hunt for life on other planets, another Sagan catch phrase applies as well: "Absence of evidence is not evidence of absence." It there once was or is now life on Mars or elsewhere in the accessible reaches of space, science must ferret it out.

Glossary

Acanothomorph. Any of diverse spiny acritarchs.

Acetic acid. An organic compound, CH_3COOH.

Acid-resistant residue. *See* Palynological maceration.

Acritarcha. An artificial taxonomic group that includes Precambrian and Phanerozoic organic-walled, commonly spheroidal, microalgal-like fossils of uncertain biologic relations.

Adaptive radiation. The evolution of a species into a group of species, adapted to different niches.

Adenine (A). One of the four nitrogenous bases in the nucleotides of DNA and RNA.

Adenosine triphosphate (ATP). A compound present in all cells that provides energy derived from sunlight (in photoautotrophs), chemical reactions (in chemoautotrophs), or food (in heterotrophs) that, stored in its phosphate bonds, can be released to power cellular processes.

Aerobe. Organism capable of using molecular oxygen (O_2) in aerobic respiration, either as a requirement to live (as in obligate aerobes) or as an energy-yielding process to supplement fermentation (as in facultative aerobes). *Also see* Complete aerobe; Facultative aerobe; Obligate aerobe.

Aerobic respiration. The oxygen-consuming, energy-yielding process carried out by almost all eukaryotes and diverse prokaryotes; "breathing."

Age of Evident Life. Informal name for the Phanerozoic Eon.

Age of Generalization and Survival. Informal name for the Precambrian Eon.

Age of Microscopic Life. Informal name for the Precambrian Eon.

Age of Specialization and Extinction. Informal name for the Phanerozoic Eon.

Alanine. One of the twenty amino acids commonly present in proteins of living systems.

Alga. Any of diverse types of eukaryotic photoautotrophic single-celled protists, such as phytoplankton or many-celled seaweeds.

ALH84001. The 1.9-kg meteorite claimed to contain evidence of past martian microbial life, named for where and when it was found (Allan Hills ice field, Antarctica, in 1984) together with its assigned sample number (001).

Alternation of generations. In sexually reproducing eukaryotes, the alternation of diploid and haploid phases of the life cycle.

Amino acid. A small molecule (monomer) containing an amino group ($-NH_2$) at one end and a carboxylic acid group ($-COOH$) at the other that can be linked to other amino acids to form a protein.

Amino group. The three-atom chemical group $-NH_2$.

Ammonia. The chemical compound NH_3.

Anaerobe. Any of various organisms, almost all prokaryotes, that can live in the absence of molecular oxygen (O_2).

Angiosperm. Any member of the taxonomic group (Angiospermae) that consists of flowering plants.

Anoxic. With reference to the complete absence of molecular oxygen (O_2).

Anoxygenic photosynthesis. The non-oxygen-producing photoautotrophy of photosynthetic bacteria.

Apex chert. A fossiliferous horizon of the Apex Basalt, a 3,465-Ma-old geologic unit of Western Australia.

Aragonite. A mineral having the same formula as calcite ($CaCO_3$) but a different crystal form and greater density.

Archaea. Any of diverse microbes of the Archaeal domain.

Archaeal domain. Together with Bacteria and Eucarya, one of three super-kingdomlike primary branches of the Tree of Life.

Archean Era. The older era of the Precambrian Eon, extending from Earth's formation 4,550 Ma ago to the beginning of the Proterozoic 2,500 Ma ago; together, the Archean and Proterozoic Eras comprise the Precambrian Eon.

Archaean. Any of the prokaryotes that belong to the Archaeal domain.

Arrested evolution. A concept essentially identical to G. G. Simpson's bradytely, proposed in 1918 by R. Ruedemann.

Asexual. *See* Nonsexual.

Aspartic acid. One of the twenty amino acids commonly present in proteins of living systems.

Asteroid. One of thousands of small planetlike bodies orbiting between Mars and Jupiter having diameters from a fraction of a kilometer to about 1,000 km.

ATP. *See* Adeosine triphosphate.

Aura seminalis. An "activator" such as subterranean heat, fluids, or percolating snow water thought in the early eighteenth century to form fossils by stimulating germination of "seeds" of animals or plants lodged in cracks and crevices of rocks.

Autotrophy. A metabolic process of plant and plantlike organisms (photoautotrophy) and diverse nonphotosynthetic bacteria and archaeans (chemoautotrophy) in which carbon dioxide serves as the principal source of cellular carbon.

Bacterial domain. Together with Archaea and Eucarya, one of three super-kingdomlike primary branches of the Tree of Life.

Bacteriochlorophyll. Any of several structurally similar light-absorbing pigments that play a central role in anoxygenic bacterial photosynthesis.

Bacterium. Any of diverse prokaryotes, including cyanobacteria, of the Bacterial domain.

Banded iron formation (BIF). Chemically deposited cherty sedimentary rock, usually thinly bedded and containing more than 15% iron.

Barite. A mineral, barium sulfate, $BaSO_4$.

Batholith. A great mass of intruded igneous rock that during its formation stopped in its rise a considerable distance below the Earth's surface.

Beta ray. An electron or positron ejected from the nucleus of an atom during radioactive decay.

Bicarbonate. A chemical, HCO_3^-, formed when carbon dioxide dissolves in water.

BIF. *See* Banded iron formation.

Biochemical. Any of a large number of chemical compounds made by a living system, composed commonly of carbon, hydrogen, oxygen, nitrogen, sulfur, and/or phosphorus (CHONSP).

Biochemistry. Chemistry that deals with the chemical compounds and processes that occur in organisms.

Biomarker. A contraction of the term "biological marker compound," used with reference to chemical fossils which by their presence show the existence of a particular biologic group (for example, fossil steranes evidencing the existence of eukaryotes).

Biosynthesis. The process of manufacture of organic compounds by biologic systems.

Biosynthetic pathway. Any of diverse enzyme-mediated multistep processes by which organic compounds are formed in living systems.

Bitter Spring Formation. A fossiliferous 850-Ma-old geologic unit of central Australia.

Bond, chemical. *See* Chemical bond.

Bond energy. Chemical energy that holds together the atoms of a molecule.

Brachiopod. Any of a phylum (Brachiopoda) of marine invertebrate animals having bivalve shells within which is a pair of small tentacle-bearing arms that bring microscopic food to the mouth; known also as lampshells.

Bradytely. An unusually slow rate (distribution) of morphological evolutionary change.

Calcareous. Containing or composed of calcium carbonate.

Calcite. A mineral, calcium carbonate, $CaCO_3$.

Cambrian explosion. The name given to the abrupt appearance of the major animal phyla in the fossil record early in the Cambrian Period.

Cambrian Period. The earliest geologic period of the Phanerozoic Eon of Earth's history, extending from 543 to 495 Ma ago.

Cambrian System. Rock strata of the Cambrian Period.

Carapace. A tough, commonly mineralized surface covering, characteristic of some animals such as crabs and lobsters.

Carbon-14 (^{14}C). A radioactive isotope of carbon produced in the upper atmosphere and present in living plants and animals that can be used in carbon-14 dating because it decays to nitrogen (^{14}N) and a beta ray with a half-life of about 5,730 years.

Carbohydrate. A carbon-, hydrogen-, and oxygen-containing organic polymeric compound composed of sugar subunits.

Carbon fixation. The metabolic incorporation of carbon into an organic compound.

Carbonaceous chondrite. A relatively rare type of stony meteorite that contains as much as 5% organic matter (kerogen) formed nonbiologically.

Carbonaceous. Containing or composed of kerogenous organic matter, whether nonbiological (as in carbonaceous chondrites) or biological (as in coal and organic-walled fossils).

Carbonate. Any of various minerals containing the chemical group CO_3^{-2}, such as calcite ($CaCO_3$) or dolomite ($CaMg [CO_3]_2$); or a rock consisting chiefly of such minerals, such as limestone or dolostone.

Carboxyl group. The four-atom chemical group -COOH.

Catalysis. Modification, especially an increase of rate, of a chemical reaction induced by a substance such as an enzyme or ribozyme that is unchanged chemically at the end of the reaction.

Cellulose. A polymeric organic compound composed of glucose subunits abundant in the walls of plant cells.

Cenozoic Era. Youngest of three eras of the Phanerozoic Eon of Earth's history, extending from the end of the Mesozoic Era, 65 Ma ago, to the present.

Cephalization. An evolutionary tendency to specialization of the body with concentration of sensory and neural organs in an anterior head.

Chalcedony. A transluscent form of quartz often deposited by chemical precipitation in vugs.

Chemical bond. A linkage between two atoms of a molecule or between atoms of neighboring molecules, often by shared electrons.

Chemoautotrophy. An autotrophic metabolism in which energy is generated by oxidation-reduction reactions of inorganic compounds.

Chert. A type of rock composed of microcrystalline quartz, SiO_2.

Chlorophyll. Any of several structurally similar light-absorbing pigments that play a central role in the oxygenic photosynthesis of cyanobacteria, algae, and higher plants.

Chloroplast. An organelle present in the cells of plants and some protists where photosynthesis occurs.

Chondrule. A particular kind of small mineralic glass spheroid present in chondritic meteorites.

CHONSP. Abbreviation for the biogenic elements: carbon, hydrogen, oxygen, nitrogen, sulfur, and phosphorus.

Chroococcaceae. A taxonomic family of simple, unicellular or colonial, spheroidal cyanobacteria.

Citric acid cycle. The electron transport cycle of aerobic respiration.

Coelom. The usually epithelium-lined space between the body wall and the digestive tract in "higher metazoans."

Coelomate. An animal that possesses a coelom.

Complete aerobe. Any of various plant and plantlike organisms, including cyanobacteria, capable both of oxygen production (by oxygenic photosynthesis) and oxygen consumption (by aerobic respiration).

Compression fossil. A flattened carbonaceous fossil preserved typically in shales and siltstones.

Condensing agent. Chemicals such as cyanogen, cyanomide, and cyanoacetylene that facilitate polymerization of monomers.

Conglomerate. A coarse-grained sedimentary rock formed of rounded fragments less than 2 mm across and cemented in a finer-grained matrix; the consolidated equivalent of gravel.

Consumer, ecologic. Any of various heterotrophs of an ecosystem.

Covalent bond. A bond formed between atoms by the sharing of electrons.

Cretaceous Period. The youngest of the three geologic periods of the Mesozoic Era of Earth's history, extending from approximately 145 to 65 Ma ago.

Cyanobacterium. Any of a diverse group (Cyanobacteria) of prokaryotic microorganisms capable of oxygen-producing photosynthesis (the group in older classifications termed blue-green algae).

Cyst. Microscopic saclike body having a robust wall that in some microalgae, protozoans, and other protists serves as a resting stage in the life cycle preceding sexual reproduction.

Cytosine (C). One of the four nitrogenous bases in the nucleotides of DNA and RNA.

Cytosol. The watery intracellular cytoplasmic fluid.

Darwinian struggle. The competition among organisms during natural selection.

Dehydration condensation. A kind of chemical reaction that involves formation of a water molecule (H-O-H) by removal of an atom of hydrogen (-H) from one monomer and a hydroxyl group (-OH) from another as the two combine into a dimer, or from one monomer and an oligomer that combine into a polymer.

Denouement. The outcome of a complex sequence of events.

Deoxyribonucleic acid (DNA). The genetic information-containing molecule of cells, a double-stranded nucleic acid made up of nucleotides that contain a nitrogenous base (adenine, guanine, thymine, or cytosine), deoxyribose sugar, and a phosphate group.

Deuterium. The hydrogen isotope that is twice the mass of ordinary hydrogen.

Development, biologic. The sequence of processes involved in growth from a fertilized egg to an adult organism.

Devonian Period. A geological period of the Phanerozoic Eon of Earth's history, extending from 410 to 360 Ma ago.

Diatom. Any of a taxonomic group of unicellular algae occurring in marine or fresh water, each having a cell wall made of two halves impregnated with silica.

Diluvialist. An adherent of the view that the biblical story of Noah and the Flood is historically accurate.

Dipeptide. An oligomer consisting of two amino acids linked by a peptide bond.

Diploid. A cell or organism possessing two sets of chromosomes such that every gene (except those governing gender) is present as two copies.

DNA. *See* Deoxyribonucleic acid.

Domain. A superkingdomlike primary branch of the Tree of Life.

Ecologic generalist. An organism capable of living in ecologically diverse habitats, such as any of many kinds of cyanobacteria.

Ecologic specialist. An organism well adapted to an ecologically limited habitat, such as any of most eukaryotes.

Ecology. The science that deals with the interrelations among organisms inhabiting a common environment and between these organisms and the environment.

Ecosystem. The complex of a biologic community and its environment viewed as an interactive unit.

Ediacaran Fauna. A latest Proterozoic-age assemblage of soft-bodied many-celled animals, the oldest fauna known.

Electron acceptor. In a chemical reaction, a molecule that accepts one or more electrons contributed by another molecule.

Electron carrier. In a chain of chemical reactions, molecules that accept electrons from an electron donor and pass these to an electron acceptor.

Electron donor. In a chemical reaction, a molecule that contributes one or more electrons to another molecule.

Endolith. Any of diverse small organisms (commonly prokaryotic, microalgal, or fungal) that live within a rock or consolidated soil crust.

Endosymbiosis. Symbiosis in which one of the partners lives within the body of the other.

Endosymbiotic hypothesis. The concept that organelles (such as mitochondria and chloroplasts) of eukaryotic cells are evolutionary descendants of free-living bacteria that established endosymbiosis with an ancestral eukaryotic host cell.

Entophysalidaceae. A taxonomic family of predominantly colonial, mucilage-enclosed elllipsoidal cyanobacteria.

Enzyme. A protein capable of catalyzing a biochemical reaction.

Eucaryal domain. Together with Bacteria and Archaea, one of three superkingdomlike primary branches of the Tree of Life.

Eukaryote. Any of a taxonomic group (the Eucarya) of organisms composed of one or more nucleus-containing cells; any member of the Eucaryal domain such as a protist, fungus, plant, or animal.

Evolution, arrested. *See* Arrested evolution.

Evolutionary distance. The degree of evolutionary relatedness between two biologic lineages.

Evolutionary stasis. Lack of evolutionary change over geologically long periods.

Exon. Any of many parts of a gene that encodes information that can be translated into protein or functional RNA.

Extremeophile. Microbes such as many archaeans that tolerate exceptionally high-temperature acidic environments.

Facultative aerobe. Any of various prokaryotes, usually bacterial, capable of aerobic respiration but that can also grow in the absence of molecular oxygen (O_2) by anaerobic metabolism.

Family, taxonomic. *See* Taxonomic family.

Fatty acid. Any of numerous monocarboxylic acids having the general formula $CnH_{2n+1}COOH$, such as acetic acid, CH_3COOH.

Fecal pellet. Organic excrement, mainly of invertebrate animals, often of simple ovoid form.

Fermentation. Anaerobic metabolism.

Ferric iron. Oxidized iron, having a valence of $+3$ as in hematite, $Fe_2^{+3}O_3^{-2}$.

Ferrous iron. Reduced iron, having a valence of $+2$ as in pyrite, $Fe^{+2}S_2^-$.

Feynman's First Principle of Science. Coined by Richard Feynman and introduced at the 1974 graduation ceremonies of the California Institute of Technology: "You must not fool yourself—and you are the easiest person [for you] to fool!"

Filament. In microbiology, a collective term for the cylindrical external sheath and cellular internal trichome of a filamentous prokaryote.

First Principle of Science. *See* Feynman's First Principle of Science.

Foraminferan. Any of a protozoal order (Foraminifera) of marine rhizopods usually having a calcareous shell perforated with minute holes for protrusion of slender pseudopodia.

Formaldehyde. An organic compound, the simplest aldehyde, HCHO, an important intermediate compound in Miller-type organic compound syntheses.

Formation, geologic. The fundamental unit in the stratigraphic classification of rocks, a definable, mappable body of igneous, sedimentary, or metamorphic rock.

Formose reaction. Chemical reaction sequence resulting in synthesis of sugars from formaldehyde; known also as the "Butlerow synthesis."

Free-living. With reference to metabolic independence, said of an organism that is neither parasitic nor symbiotic.

"Fubarized." An informal term meaning "fouled up beyond all recognition."

Gamete. Haploid female (egg) and male (sperm) sex cells, in animals formed by meiosis and in plants by mitotic division of haploid cells derived from meiotically produced spores.

Gas chromatography-mass spectrometry (GCMS). A technique that can be used to isolate and identify organic compounds.

GCMS. *See* Gas chromatography–mass spectrometry.

Gene. A segment of DNA containing information for production of a protein or RNA (ribonucleic acid) molecule.

Gene duplication. The process by which multiple copies of genes are biosynthesized.

Generalist, ecologic. *See* Ecologic generalist.

Genus. In biological classification, a major category ranking above the species and below the family.

Geotype. The genetic constitution of an organism, in contrast to the physical (including biochemical) products of gene expression (the phenotype).

Glucose. A six-carbon sugar, $C_6H_{12}O_6$, the universal fuel of life; often abbreviated CH_2O.

Glucose biosynthesis. A multistep biosynthetic pathway in which glucose is made from pyruvate.

Glycine. One of the twenty amino acids commonly present in proteins of living systems.

Glycolysis. A type of fermentation in which each molecule of glucose is broken down to produce energy and two molecules of pyruvate.

Gondwanaland. The late Paleozoic to early Mesozoic supercontinent composed of the land masses of South America, Africa, Madagascar, India, and Antarctica.

Gram-negative bacteria. An informal taxonomic group composed of bacteria having cell walls that do not retain Gram's stain.

Gram-positive bacteria. An informal taxonomic group composed of bacteria having cell walls that retain Gram's stain.

Graphite. A mineral, crystalline carbon, C_6.

Greenhouse effect. A warming of the Earth's surface and lower layers of the atmosphere caused by interaction of solar radiation with atmospheric gases (mainly carbon dioxide, methane, and water vapor) and its conversion to heat.

Greenhouse effect, reversed. *See* Reversed greenhouse effect.

Greenhouse gas. Gases such as carbon dioxide, methane, and water vapor that produce a greenhouse effect.

Greenstone belt. A thick pile of interlayered volcanic and sedimentary rocks, typical especially of Archean geologic terrains.

Green sulfur bacteria. A taxonomic group of bacteria that includes anoxygenic photosynthesizers such as *Chlorobium*.

Group, geologic. A major category in the stratigraphic classification of rocks ranking above the formation and below the supergroup.

Group, stromatolite. *See* Stromatolite group.

Growth surface. The uppermost accretionary layer of a stromatolite, comprised typically of various kinds of cyanobacteria.

Guanine (G). One of the four nitrogenous bases in the nucleotides of DNA and RNA.

Gunflint Formation. A fossiliferous 2,100-Ma-old geologic unit of southern Ontario, Canada.

Gymnosperm. Any member of the taxonomic group (Gymnospermae) that consists of plants having naked seeds such as conifers, cycads, and *Ginkgo*.

Gypsum. A mineral, hydrous calcium sulfate, $CaSO_4 \cdot H_2O$.

Half-life. In geochronology, the time required for half of the atoms of an amount of a radioactive isotope to be transformed to daughter atoms.

Haploid. In sexually reproducing organisms, the chromosomal complement present in sperm or egg.

Hematite. A mineral, the iron oxide Fe_2O_3.

Heredity. The transmission of genetic factors that determine individual characteristics from one generation to the next.

Heterotrophic hypothesis. The concept introduced by A. I. Oparin and J.B.S. Bernal that the earliest forms of life were heterotrophs that used nonbiologically produced organic matter as their carbon source.

Heterotrophy. The metabolic process of animals and animal-like organisms in which organic compounds serve as the source of carbon and energy.

Horotely. The rate (distribution) of morphological evolutionary change shown by most Phanerozoic organisms.

Hydrocarbon. Any of a diverse group of organic compounds composed of hydrogen and carbon.

Hydrogen cyanide. An organic compound, HCN.

Hydrogen source. A chemical source of hydrogen atoms (electrons) that during photosynthesis combine with carbon dioxide to form glucose sugar.

Hydrogen sulfide. A gaseous organic compound, H_2S.

Hydrolysis. A process of breakdown of chemical compounds that involves addition of the chemical elements of water (HOH) to the substance broken down.

Hydrophilic. Having affinity for water.

Hydrophobic. Lacking affinity for water.

Hydroxyl group. The two-atom chemical group -OH.

Hyellaceae. A taxonomic family of predominantly endolithic cyanobacteria.

Hyperthermophile. Any of various prokaryotic microorganisms, such as diverse members of the Archaea, that survive and grow in exceptionally high temperature ($>80°$ C) environments.

Hypobradytely. The exceptionally slow rate (distribution) of morphological evolutionary change characteristic of cyanobacteria.

IDP. *See* Interplanetary dust particle.

Igneous. Rock formed from solidification of molten magma.

Index fossil. Any of diverse fossils that—because they are easily recognizable, exist over a relatively short period of geologic time, and are widespread—can be used to determine the age of rock strata in which they are preserved.

Indigenous. Having originated in a particular setting.

Industrial Revolution. A rapid major change in an economy, as in England beginning in the late eighteenth century, marked by the general introduction of power-driven machinery.

Infrared light. That part of the electromagnetic spectrum lying outside the visible range at its red end.

Intermediate compound. Any chemical compound formed in a reaction series prior to formation of the final product.

Interplanetary dust particle (IDP). Small, often microscopic rocky particles of interplanetary debris.

Invertebrate. Any of diverse animals that lack backbones; "lower metazoans."

Ion. An electrically charged atom or group of atoms.

Isochron. A parameter used in isotopic dating of geologic materials, experimentally determined from comparison of the isotopic compositions of two or more components (usually minerals) that share a common age.

Isotope. Any of two or more types of atoms of a chemical element that have nearly identical chemical behavior but differing atomic mass and physical properties, whether stable or subject to radioactive decay.

Isotopic date. Age of a rock (or organic substance less than 60,000 years old) determined by measurement of the ratio of a parent isotope to one of the products of its radioactive decay.

Isotopic fractionation. Separation of isotopes of an element, in organisms often mediated by an enzyme.

Kerogen. Particulate, geochemically altered, macromolecular organic matter, insoluble in organic solvents and mineral acids, present in sedimentary rocks and carbonaceous meteorites.

Kilocalorie. A unit of energy, the amount of heat required to raise the temperature of 1 kg of water 1° C.

Kinetic isotopic fractionation. Separation of isotopes of an element as a result of their speeds of movement, in organisms mediated by an enzyme that interacts more readily with one of two or more isotopes of an element.

Lactic acid. An organic acid, $C_3H_6O_3$, often produced by fermentation.

Lava. Fluid rock that issues from a volcano.

Law of Superposition. The geologic rule that, in a sequence of undeformed strata, the oldest rock layer is at the bottom and each layer upward is progressively younger.

Light-harvesting pigment. Organic compounds, such as chlorophyll and bacteriochlorophyll, that absorb light energy in photosynthesis.

Limestone. A kind of sedimentary rock consisting mainly of calcium carbonate minerals.

Lungfish. Any of six living and numerous fossil species that, together with the coelacanth, are classified as lobe-finned fish (which have both bone and muscle in their limbs, in contrast to the simple fins of bony, "teleost" fish).

Ma. Mega anna, one million (1×10^6) years.

Maceration. *See* Palynological maceration.

Magma. Molten rock such as volcanic lava.

Magnetite. A mineral, the iron oxide Fe_3O_4.

Magnetofossil. Minute grains of the mineral magnetite present in sedimentary rocks suspected to be of bacterial origin.

Mass spectrometry. Instrumental method of identifying the chemical constituents of a substance by means of the separation of gaseous ions according to their differing mass and charge.

Megasphaeromorph. Spheroidal acritarchs larger than 200 μm in diameter.

Meiosis. The process of nuclear division that reduces the number of chromosomes from the diploid to the haploid number in each of four product cells (in animals, the sperm or egg).

Mesosphaeromorph. Spheroidal acritarchs 60 to 200 μm in diameter.

Mesozoic Era. The second oldest of three geologic eras of the Phanerozoic Eon of Earth's history, extending from the end of the Paleozoic Era, 251 Ma ago, to the beginning of the Cenozoic Era, 65 Ma ago.

Metabolic pathway. Any of several enzyme-mediated multistep processes by which metabolism is carried out in living systems.

Metabolism. The sum of the energy-consuming and energy-producing processes that happen during the chemical buildup and breakdown of organic compounds in living systems.

Metamorphic rock. Rock derived as a result of mineralogical, chemical, and structural changes in a preexisting rock in response to changes in temperature, pressure, and chemical environment at depth in the Earth's crust.

Metamorphism. A change in the constitution of a rock produced by pressure and heat, "geologic pressure cooking."

Metazoan. A taxonomic category comprised of many-celled animals.

Meteor. Solar system matter observable when it falls through Earth's atmosphere and is heated by friction to temporary incandescence; a "shooting star."

Meteorite. A meteor that reaches Earth's surface.

Methane. The colorless gaseous hydrocarbon CH_4.

Methane-generating prokaryote. Any various archaeal microbes that give off methane gas as a product of their chemoautotrophic metabolism.

Methanogen. *See* Methane-generating prokaryote.

Methyl group. The four-atom chemical group $-CH_3$.

Microalga. Any of diverse microscopic photoautotrophic protists, especially unicellular eukaryotic phytoplankton.

Microbe. Informal term for any of diverse types of prokaryotic bacteria or archaeans.

Micrometer (μm). A unit of length, one-millionth (10^{-6}) of a meter.

Mitochondrion. An organelle of eukaryotic cells, the site of aerobic respiration.

Mitosis. A type of division of the cell nucleus resulting in formation of two daughter cells, each a genetic copy (clone) of the parent cell; in unicellular eukaryotes, a type of nonsexual reproduction.

Mitotic spindle. Protein-rich fibers that separate paired chromosomes during mitosis, pulling them to opposite ends of the dividing cell.

Mollusk. Any of an animal phylum (Mollusca) characterized by a large muscular foot and a mantle that secretes spicules or shells, such as a snail, clam, or squid.

Monomer. A chemical compound, usually small, that can be linked to other monomers into a larger, multicomponent polymer.

Mononucleotide. A nucleotide composed of one monomer each of a nitrogenous base, a sugar, and a phosphoric acid.

Monosaccharide. A sugar monomer.

Morphometric. Pertaining to measurement of external form.

Mutation. Any change in the nucleotide sequence of a gene.

Mycoplasma. Any of a genus (*Mycoplasma*) of extremely minute parasitic microorganisms that lack cell walls; known also as PPLO, *Pleuropneumonia*-like organisms.

Nanometer (nm). A unit of length, one billionth (10^{-9}) of a meter.

Natural selection. The preferential survival of individuals having advantageous variations relative to other members of their population or species; for natural selection to operate, there must be competition for resources (a struggle for survival) and suitable variation among individuals.

Necessities of life. A "carbon source" (that is, a source of CHONSP) and a source of energy.

Nematode. Any of a particular phylum (Nematoda) of small cylindrical worms.

Neo-Darwinian Synthesis. A merging of the concepts of Darwinian selection, Mendelian genetics, random mutation, population biology, and paleontology in the 1930s and 1940s that resulted in the predominating theory that evolution is change in the genetics of species due to natural selection.

Newt. Any of various small semiaquatic salamanders.

Niche, ecologic. A unique way of life for a species of organism.

Nif **complex.** *See* Nitrogenase enzyme complex.

Nitrate. NO_3^-, a biologically usable form of nitrogen.

Nitrogenase enzyme complex. The enzyme complex involved in biological nitrogen fixation.

Nitrogen fixation. The biologic process carried out by various bacterial and archaeal prokaryotes in which molecular nitrogen (N_2) is combined with hydrogen to produce ammonia, a biologically useable form of nitrogen.

Nonsexual. With reference to organisms that lack capability to reproduce sexually.

Nucleic acid. The genetic information-containing organic acids DNA and RNA.

Nucleoside. A chemical compound consisting of a purine or pyrimidine nitrogenous base combined with deoxyribose or ribose sugar.

Nucleotide. Any of several compounds that consist of a sugar (deoxyribose or ribose) linked to a purine (adenine [A] or guanine [G]) or a pyrimidine (thymine [T], cytosine [C], or uracil [U]) nitrogenous base and a phosphate group, the basic structural units of DNA and RNA.

Nucleus. In eukaryotes, a membrane-enclosed organelle that contains the chromosomes.

Obligate aerobe. An organism unable to live in the absence of molecular oxygen (O_2), needed to support its aerobic respiration.

Obligate anaerobe. An organism incapable of growth and reproduction in the presence of molecular oxygen (O_2).

Oligomer. Any small polymeric chemical compound composed of a few or several monomeric units.

Ordovician Period. A geological period of the Phanerozoic Eon of Earth's history, extending from 495 to 440 Ma ago.

Organelle. Any of various specialized membrane-enclosed structures in a cell, such as mitochondria or plastids, that perform a specific function.

Organic chemistry. Chemistry that deals with organic compounds, including but not limited to chemical compounds and processes that occur in organisms.

Organic compound. A chemical compound of the type typical of (but not restricted to) living systems, composed commonly of carbon, hydrogen, oxygen, nitrogen, sulfur, and/or phosphorus.

Oscillatoriaceae. A taxonomic family of simple filamentous cyanobacteria that lack heterocysts, a particular kind of specialized cell.

Oxic. Pertaining to the presence of molecular oxygen (O_2).

Oxidation. A chemical process in which oxygen combines with another element (or in which hydrogen or electrons are removed from an element) and energy in the form of heat is released.

Oxide. A chemical compound of oxygen and another element, as in the minerals hematite, Fe_2O_3, and magnetite, Fe_3O_4.

Oxygenic photosynthesis. Oxygen-producing photoautotrophy such as that in plants and plantlike organisms, including cyanobacteria.

Oxygen sink. Any of several molecular oxygen-consuming processes (such as aerobic respiration) or substances (such as unoxidized volcanic gases or dissolved ferrous iron).

Ozone. A triatomic form of oxygen, O_3, formed naturally in the upper atmosphere.

PAH. *See* Polycyclic aromatic hydrocarbon.

Paleosol. A geologically preserved soil horizon.

Paleozoic Era. The oldest of three geologic eras of the Phanerozoic Eon of Earth's history, extending from the end of the Proterozoic Era of the Precambrian Eon, 543 Ma ago, to the beginning of the Mesozoic Era, 251 Ma ago.

Palynological maceration. An organic (usually fossil-containing) residue prepared by dissolution of a sedimentary rock in hydrochloric, hydrofluoric, or other mineral acid.

Palynology. The scientific study of plant pollen and spores, especially of fossil plants.

Paradigm. A well-established example, an archetype.

Peptide. A chemical compound, such as a protein, in which components are linked by peptide bonds.

Peptide bond. The type of chemical bond that links adjacent amino acids in proteins.

Period, geologic. A formal division of geologic time longer than an epoch and included in an era.

Permineralize. *See* Petrify.

Petrify. The process by which organic-walled organisms can be preserved as fossils, embedded three-dimensionally in a siliceous, calcitic, or other mineral matrix; synonymous with permineralize.

Petrographic thin section. A slice of rock ground sufficiently thin that light can be transmitted through it.

Phanerozoic Eon. The younger of two principal divisions (eons) of Earth's history, extending from the beginning of the Cambrian Period, 543 Ma ago to the present; the Phanerozoic and the older Precambrian Eon comprise all of geologic time.

Phenotype. The physical (including biochemical) products of gene expression of an organism, in contrast to its genetic constitution (the genotype).

Phosphoglyceric acid. Phosphoglycerate, an important intermediate compound in the degradative and biosynthetic pathways of glucose.

Photic zone. The surface layers of oceans or lakes where sufficient light penetrates to support photosynthesis.

Photoautotrophy. Autotrophy powered by light energy.

Photodissociation. The breaking apart of a chemical compound by photolysis.

Photoheterotrophy. The metabolic process in which light energy is used to aid the uptake into cells of exogenous organic compounds.

Photolysis. Decomposition of a chemical compound by the action of light energy.

Photosynthesis. The metabolic process carried out by photosynthetic bacteria, cyanobacteria, algae, and plants in which light energy is converted to chemical energy and stored in molecules of biosynthesized carbohydrates.

Photosynthetic bacterium. Any of diverse types of bacteria capable of anoxygenic photosynthesis.

Photosynthetic space. The surface area or volume within the photic zone where photosynthesis can occur.

Phototaxis. Movement of an organism in response to a gradient of light.

Plankton. Organisms inhabiting the surface layers of a sea or lake, such as small drifting algae, protozoans, and animals.

Plate tectonics. Global tectonics based on an Earth model characterized by many thick oceanic or continental plates that move slowly across the global surface propelled by movement of underlying material of the planetary interior.

Pleurocapsaceae. A taxonomic family of predominantly colonial, mucilage-enclosed ellipsoidal cyanobacteria.

Polycyclic aromatic hydrocarbon (PAH). Any of various organic compounds composed of a few to many six-membered rings of carbon atoms linked by an alternating sequence of single and double bonds and to which hydrogen atoms are attached.

Polymer. A multicomponent chemical compound, usually large, that consists of many monomers linked together.

Polymerase. An enzyme that builds polymers from monomeric subunits.

Polynucleotide. Any of various nucleic acids, polymers of nucleotides.

Polypeptide. Any of various proteins, polymers of amino acids.

Polysaccharide. Any of various carbohydrates, polymers of monosaccharides.

Population. Any naturally occurring group of organisms of the same species that occupies a relatively well-defined geographic region and has reproductive continuity from generation to generation.

Precambrian-Cambrian boundary problem. The series of interrelated scientific questions relating to the first appearance of diverse shelled metazoans at the beginning of the Cambrian Period of Earth's history.

Precambrian Eon. The older of two principal divisions (eons) of Earth's history, extending from the formation of the planet, 4,550 Ma ago, to the beginning of the Cambrian Period, 543 Ma ago; the Precambrian and the younger Phanerozoic Eon comprise all geologic time.

Principle of Conservatism and Economy. The rule that in evolution new structures and functions arise through modification of previously established systems.

Prokaryote. Any of diverse types of nonnucleated microorganisms of the Archaea and Bacteria.

Protein. A polymeric organic compound composed of amino acid monomers.

Proterozoic Era. The younger era of the Precambrian Eon, extending from the end of the Archean Era, 2,500 Ma ago, to the beginning of the earliest (Cambrian) period of the Phanerozoic Eon, 543 Ma ago; together, the Proterozoic and the Archean Eras comprise the Precambrian Eon.

Protist. A general term for unicellular plant- and animal-like eukaryotes.

Protozoan. Animal-like protists.

Purine. A type of nitrogenous base in nucleic acid, such as adenine or guanine.

Purple bacteria. A taxonomic group of bacteria that includes anoxygenic photosynthesizers such as *Chromatium.*

Pylome. A commonly circular porelike opening in the wall of an acritarch that functions in release of reproductive bodies during excystment.

Pyrimidine. A type of nitrogenous base in nucleic acid, such as cytosine, thymine, or uracil.

Pyrite. A mineral, iron sulfide, FeS_2; "fool's gold."

Pyruvate. A three-carbon organic compound formed by glycolytic breakdown of the six-carbon sugar, glucose.

Quartz. A mineral, silicon dioxide, SiO_2.

Radioactivity. The property possessed by isotopes of some elements (such as uranium) of spontaneously emitting alpha or beta rays by the disintegration of the nuclei of atoms.

Reduction. In chemistry, to combine with or subject to the action of hydrogen, thereby lowering the state of oxidation.

Reversed greenhouse effect. A cooling of the Earth's surface and lower layers of the atmosphere caused by removal of atmospheric greenhouse gases, principally carbon dioxide.

Ribonucleic acid (RNA). A single-stranded nucleic acid made up of nucleotides that contain a backbone of ribose sugar and phosphate to which are linked the nitrogenous bases adenine, uracil, guanine, and cytosine.

Ribose. An organic compound, a five-carbon sugar of the formula $C_5H_{10}O_5$.

Ribosomal ribonucleic acid (rRNA). Any of several RNAs present in ribosomes.

Ribosome. The intracellular body in eukaryotes and prokaryotes where protein synthesis occurs.

Ribozyme. Any of several kinds of RNAs that have enzymelike properties.

Ribulose bisphosphate carboxylase/oxygenase. *See* RUBISCO.

RNA. *See* Ribonucleic acid.

Rosetta Stone. A slab of black basalt stone found in 1799 that bears an inscription in hieroglyphics, demotic characters, and Greek, celebrated for having given the first clue to the decipherment of Egyptian hieroglyphics; from this, any breakthrough discovery of great magnitude.

rRNA. *See* Ribosomal ribonucleic acid.

RUBISCO. The carbon dioxide-fixing enzyme of cyanobacteria, algae, and plants, ribulose bisphosphate carboxylase/oxygenase.

Rule of the Survival of the Relatively Unspecialized. A rule coined by G. G. Simpson that because of a lack of ecologic flexibility, more specialized organisms tend to become extinct before less specialized ones.

Sclerite. A physically resistant plate or spicule that forms part of an animal exoskeleton.

Sediment. Particulate, commonly granular mineralic material deposited by water, wind, or glaciers that on compression can be lithified to a sedimentary rock.

Seed plant. Any of diverse "higher land plants" such as gymnosperms and angiosperms that produce seeds.

Serpentine. A metamorphic mineral usually having a dull green color and often mottled appearance, the hydrous magnesium silicate, $Mg_3Si_2O_7 \cdot 2H_2O$.

Sexual reproduction. In eukaryotes, the process of reproduction involving formation of spores (in plants) and gametes (in animals) by meiosis, followed by fusion of gametes (syngamy).

Shale. A sedimentary rock formed by consolidation of clay or mud.

Sheath. The tubular extracellular mucilage surrounding the cellular trichome of a filamentous prokaryote.

Silica. A chemical compound, silicon dioxide, SiO_2.

Siltstone. A sedimentary rock formed by consolidation of fine silt.

Silurian Period. A geological period of the Phanerozoic Eon of Earth's history, extending from 440 to 410 Ma ago.

Specialist, ecologic. *See* Ecologic specialist.

Species. The fundamental category of biological classification, ranking below the genus and in some species composed of subspecies or varieties; of various definitions, the most common is the "Biological Species Concept": "Species are actually or potentially interbreeding natural populations which are reproductively isolated from other such groups."

Spermatick Principle. The seeds or eggs of plants or animals lodged in cracks and crevices in rocks thought in the seventeenth century to germinate into fossils by action of an extraordinary "life-giving force."

Sphaeromorph. Any of various types of morphologically simple spheroidal acritarchs.

Spore. The haploid product of meiosis in plants.

Spore plant. Any of various "lower land plants" that instead of producing seeds (as do gymnosperms and angiosperms) reproduce by shedding spores, such as club mosses (lycophytes) and horse tails (sphenophytes).

Sports of nature. Fossil-like mineralic objects formed by nonbiologic processes.

Stasis, evolutionary. *See* Evolutionary stasis.

Sterane. Any of a group of organic compounds derived by geochemical hydrogenation of sterols.

Sterol. Any of various cyclic steroid alcohols, such as cholesterol ($C_{27}H_{45}OH$), present especially in the lipid membranes of eukaryotes.

Strategies of life. Autotrophy and heterotrophy, strategies evolved to meet the necessities of life.

Stratum. A layer or horizon of sedimentary rock.

Strecker synthesis. A chemical reaction series by which amino acids are formed from aldehyde, hydrogen cyanide, and ammonium.

Stromatolite. An accretionary organosedimentary structure, commonly finely layered, megascopic, and calcareous, produced by the activities of mat-building microorganisms, principally filamentous photosynthetic prokaryotes such as various types of cyanobacteria.

Stromatolite group. A taxonomic category used in classification of stromatolites and more or less equivalent to a paleontologic "form genus."

Succus lapidenscens. A "lapidifying juice" thought in the sixteenth and seventeenth centuries to transform rocks into superficially lifelike objects, thereby explaining the origin of fossils.

Sugar. Any of various monosaccharides having the generalized formula CH_2O, such as fructose, sucrose, and glucose.

Sulfate-reducing bacterium. Any of various chemoautotrophic bacteria that derive energy from reduction of sulfate (SO_4^{-2}), such as *Desulfovibrio*.

Sulfide. A chemical compound of sulfur and another element, as in the mineral pyrite, FeS_2, or gaseous hydrogen sulfide, H_2S.

Supergroup, geologic. A major category in the stratigraphic classification of rocks ranking above the group.

Symbiosis. The living together in mutually dependent close association of two often taxonomically dissimilar organisms.

Syngamy. The fusion of gametes (egg and sperm) in sexual reproduction.

Syngenetic. Of the same origin, formed at the same time.

System, geologic. The term applied to the rocks of a specified geologic period; for example, rocks formed during the Cambrian Period comprise the Cambrian System.

Tachytely. An unusually fast rate (distribution) of morphological evolutionary change.

Taxon. In biologic classification, pertaining to a unit of any rank (that is, a particular species, genus, family, class, order, or division or phylum) or the scientific name of that unit.

Taxonomic family. In biologic classification, a major category ranking above the genus and below the class.

Taxonomic occurrence. The number of taxa of a particular rank known to be present in named geologic units of a specified kind (for example, the number of species of a particular group reported from one or more geologic formations).

Taxonomy. The theory and practice of classifying organisms.

Tectonic. Pertaining to a crust-deforming process or event.

Thermophile. Any of various organisms, such as diverse prokaryotes, that can survive and grow in relatively high-temperature environments such as hot springs.

Thin section. *See* Petrographic thin section.

Thymine (T). One of the four nitrogenous bases in the nucleotides of DNA.

Trace fossil. A type of metazoan fossil consisting of a track, trail, burrow, or similar structure.

Tree of Life. A branching, treelike representation showing the relatedness of all living organisms, commonly based on comparison of rRNAs, the ribonucleic acids of protein-manufacturing ribosomes.

Trichome. The living cellular part of a sheath-enclosed microbial filament.

Trilobite. Any of a group (Trilobita) of extinct arthropod animals of the Paleozoic Era (543 to 245 Ma ago), characterized by a three-lobed body organization.

Ultraviolet (UV) light. That part of the electromagnetic spectrum lying outside the visible range at its violet end.

Undermat. Layers of a stromatolite immediately below the growth surface, comprised typically of photosynthetic bacteria.

Universal Tree of Life. *See* Tree of Life.

Uracil (U). One of the four nitrogenous bases in the nucleotides of RNA.

Uraninite. A mineral, uranium oxide, approximately of the composition UO_2.

UV light. *See* Ultraviolet (UV) light.

Vertebrate. Any member of the subphylum Vertebrata that consists of all animals that possess a bony or cartilaginous skeleton and a well-developed brain, such as fishes, amphibians, reptiles, birds, and mammals.

Vesicle. In planktonic cyanobacteria and other prokaryotes, a gas-filled pocket used to control buoyancy.

Visible light. That part of the electromagnetic spectrum visible to humans and lying between the infrared and ultraviolet parts of the spectrum.

Volkswagen Syndrome. A state in which the unchanging external form of a set of objects over time masks significant internal changes (named after the Volkswagen "beetle" automobiles of the 1950s to 1970s, known for their more or less constant body form that masked significant evolution of their internal machinery).

Vug. A small unfilled cavity in a rock.

Weathering. The chemical and physical processes that disaggregate a rock into its component mineral grains or crystals.

Yeast. A minute fungus, *Saccharomyces cerevisiae.*

Zircon. A mineral, the zirconium silicate $ZrSiO_4$.

Zygote. The diploid cell formed by fusion (syngamy) of two haploid gametes, the earliest formed cell of the embryo of animals or the spore-producing generation of plants.

Further Reading

Chapter 1

Cloud, P. 1983. Early biogeologic history: The emergence of a paradigm. In J. W. Schopf, ed., *Earth's Earliest Biosphere: Its Origin and Evolution*, 14–31. Princeton, N.J.: Princeton University Press.

Darwin, C. *On the Origin of Species by Means of Natural Selection, or the Preservation of Favoured Races in the Struggle for Life*. Facsimile of the 1st edition of 1859, Cambridge, MA: Harvard University Press, 1964. Facsimile of the 6th (and last) edition of 1872, London: John Murray, 1902.

Dawson, J. W. 1875. *The Dawn of Life*. London: Hodder and Stoughton.

Gould, S. J. 1989. *Wonderful Life*. New York: W. W. Norton.

O'Brien, C. F. 1971. Sir William Dawson, a life in science and religion. *Memoirs of the American Philosophical Society* 84: 1–207.

O'Brien, C. F. 1970. *Eozoön Canadense,* "the dawn animal of Canada." *Isis* 61: 206–223.

Schopf, J. W. 1992. Historical development of Proterozoic micropaleontology. In J. W. Schopf, and C. Klein, Eds., *The Proterozoic Biosphere: A Multidisciplinary Study*, 179–183. New York: Cambridge University Press.

Seward, A. C. 1931. *Plant Life through the Ages*. Cambridge, U.K.: Cambridge University Press.

Yochelson, E. L. 1997. *Charles Doolittle Walcott, Paleontologist*. Kent, Ohio: Kent State University Press.

Chapter 2

Cloud, P. 1972. A working model for the primitive Earth. *American Journal of Science* 272: 537–548.

Cloud, P. 1983. Early biogeologic history: The emergence of a paradigm. In J. W. Schopf, ed., *Earth's Earliest Biosphere: Its Origin and Evolution*, 14–31. Princeton, N.J.: Princeton University Press.

Radhakrishna, B. P., ed. 1991. *The World of Martin F. Glaessner*, iii–xxiv. Memoir No. 20. Bangalore, India: Geological Society of India.

Schopf, J. W. 1992. Historical development of Proterozoic micropaleontology. In J. W. Schopf and C. Klein, eds., *The Proterozoic Biosphere: A Multidisciplinary Study*, 179–183. New York: Cambridge University Press.

Chapter 3

Blake, T. S., and McNaughton, N. J. 1984. A geochronological framework for the Pilbara region. In J. R. Muhling, D. K. Groves, and T. S. Blake, eds., University of Western Australia, Geology Department and University Extension, Publication 9.

Groves, D. I., Dunlop, J.S.R., and Buick, R. 1981. An early habitat of life. *Scientific American* 245: 64–73.

Schopf, J. W. 1992. Paleobiology of the Archean. In J. W. Schopf and C. Klein, eds., *The Proterozoic Biosphere: A Multidisciplinary Study*, 25–39. New York: Cambridge University Press.

Schopf, J. W. 1993. Microfossils of the Early Archean Apex chert: New evidence of the antiquity of life. *Science* 260: 640–646.

Schopf, J. W. and Walter, M. R. 1983. Archean microfossils: New Evidence of ancient microbes. In J. W. Schopf, Ed., *Earth's Earliest Biosphere: Its Origin and Evolution*, 214–239. Princeton, N.J.: Princeton University Press.

Chapter 4

Iwabe, N., Kuma, K., Hasegawa, M., Osawa, S., and Miyata, T. 1989. Evolutionary relationships of archaebacteria, eubacteria and eukaryotes inferred from phylogenetic trees of duplicated genes. *Proceedings of the National Academy of Sciences USA* 86: 9355–9359.

Miller, S. L. 1953. A production of amino acids under possible primitive Earth conditions. *Science* 117: 528–529.

Miller, S. L. 1974. The first laboratory synthesis of organic compounds under primitive Earth conditions. In J. Neyman, ed., *The Heritage of Copernicus: Theories "Pleasing to the Mind,"* 228–242. Cambridge, Mass.: MIT Press.

Olsen, G. J., and Woese, C. R. 1993. Ribosomal RNA: A key to phylogeny. *FASEB Journal* 7: 113–123.

Oparin, A. I. 1924. *Proiskhozhdenie Zhiznii* [*The Origin of Life*]. Moscow: Moskovskii Rabochii. English translation in J. D. Bernal, *The Origin of Life*. London: Weidenfeld and Nicolson, 1967.

Oparin, A. I. 1938. *The Origin of Life*. New York: Macmillan.

Urey, H. C. 1952. On the early chemical history of the Earth and the origin of life. *Proceedings of the National Academy of Sciences USA* 38: 351–363.

Woese, C. R. 1987. Bacterial evolution. *Microbiological Reviews* 51: 221–271.

Woese, C. R., Kandler, O., and Whellis, M. L. 1990. Towards a natural system of organisms: Proposal for the domains Archaea, Bacteria, and Eucarya. *Proceedings of the National Academy of Sciences USA* 87: 4576–4579.

Chapter 5

Blankenship, R. E. 1992. Origin and early evolution of photosynthesis. *Photosynthesis Research* 33: 91–111.

Broda, E. 1975. *The Evolution of Bioenergetic Processes*. New York: Pergamon Press.

Cech, T. 1986. RNA as an enzyme. *Scientific American* 255 (5): 64–75.

Eck, R. V., and Dayhoff, M. O. 1966. Evolution of the structure of ferredoxin based on living relics of primitive amino acid sequences. *Science* 152: 363–366.

Schopf, J. W. 1992. Times of origin and earliest evidence of major biologic groups. In J. W. Schopf and C. Klein, eds., *The Proterozoic Biosphere: A Multidisciplinary Study*, 587–593. New York: Cambridge University Press.

Schopf, J. W. 1995. Metabolic memories of Earth's earliest biosphere. In C. R. Marshall and J. W. Schopf, eds., *Evolution and the Molecular Revolution*, 73–107. Boston: Jones and Bartlett.

Sonti, R. V., and Roth, J. R. 1989. Role of gene duplications in the adaptation of *Salmonella typhimurium* to growth on limiting carbon sources. *Genetics* 123: 19–28.

Woese, C. R. 1987. Bacterial evolution. *Microbiological Reviews* 51: 221–271.

Woese, C. R., Kandler, O., and Wheelis, M. L. 1990. Towards a natural system of organisms: Proposal for the domains Archaea, Bacteria, and Eucarya. *Proceedings of the National Academy of Sciences USA* 87: 4576–4579.

Chapter 6

Cloud, P. 1976. Beginnings of biospheric evolution and their biogeochemical consequences. *Paleobiology* 2: 351–387.

Hayes, J. M. 1994. Global methanotrophy at the Archean-Proterozoic transition. In S. Bengtson, ed., *Early Life on Earth*, 220–236. New York: Columbia University Press.

Holland, H. D. 1994. Early Proterozoic atmospheric change. In S. Bengtson, ed., *Early Life on Earth*, 237–244. New York: Columbia University Press.

Klein, C., and Buekes, N. J. 1992. Time distribution, stratigraphy, sedimentologic setting, and geochemistry of Precambrian iron-formations. In J. W. Schopf and C. Klein, eds., *The Proterozoic Biosphere: A Multidisciplinary Study*, 139–146. New York: Cambridge University Press.

Schidlowski, M., Hayes, J. M., and Kaplan, I. R. 1983. Isotopic inferences of ancient biochemistries: Carbon, sulfur, hydrogen, and nitrogen. In J. W. Schopf, ed., *Earth's Earliest Biosphere: Its Origin and Evolution*, 149–186. Princeton, N.J.: Princeton University Press.

Sleep, N. H., Zahnle, K. J., Kasting, J. F., and Morowitz, H. J. 1989. Annihilation of ecosystems by large asteroid impacts on the early Earth. *Nature* 342: 139–142.

Strauss, H., Des Marais, D. J., Hayes, J. M. and Summons, R. E. 1992. The carbon-isotopic record. In J. W. Schopf and C. Klein, eds., *The Proterozoic Biosphere: A Multidisciplinary Study*, 117–127. New York: Cambridge University Press.

Chapter 7

Monastersky, R. 1998. The rise of life on Earth. *National Geographic* 193(3): 54–81.

Pierson, B. K., Bauld, J., Castenholz, R. W., D'Amelio, E., Des Marais, D. J., Farmer, J. D., Grotzinger, J. P., Jørgensen, B. B., Nelson, D. C., Palmisano, A. C., Schopf, J. W., Summons, R. E., Walter, M. R., and Ward, D. M. 1992. Modern mat-building microbial communities: A key to the interpretation of Proterozoic stromatolitic communities. In J. W. Schopf and C. Klein, eds., *The Proterozoic Biosphere: A Multidisciplinary Study*, 245–342. New York: Cambridge University Press.

Chapter 8

Golubic, S. 1976. Organisms that build stromatolites. In M. R. Walter, ed., *Stromatolites, Developments in Sedimentology* 20, 113–126. Amsterdam: Elsevier.

Golubic, S., and Hofmann, H. J. 1976. Comparison of Holocene and mid-Precambrian Entophysalidaceae (Cyanophyta) in stromatolitic mats: Cell division and degradation. *Journal of Paleontology* 50: 1074–1082.

Mendelson, C. V., and Schopf, J. W. 1992. Proterozoic and selected Early Cambrian microfossils and microfossil-like objects. In J. W. Schopf and C. Klein, eds., *The Proterozoic Biosphere: A Multidisciplinary Study*, 865–951. New York: Cambridge University Press.

Ruedemann, R. 1918. The paleontology of arrested evolution. *New York State Museum Bulletin* 196: 107–134.

Ruedemann, R. 1922a. Additional studies of arrested evolution. *Proceedings of the National Academy of Sciences USA* 8: 54–55.

Ruedemann, R. 1922b. Further notes on the paleontology of arrested evolution. *American Naturalist* 56: 256–272.

Schopf, J. W. 1987. "Hypobradytely": Comparison of rates of Precambrian and Phanerozoic evolution. *Journal of Vertebrate Paleontology* 7 (3, suppl.): 25.

Schopf, J. W. 1992. Tempo and mode of Proterozoic evolution. In J. W. Schopf and C. Klein, eds., *The Proterozoic Biosphere: A Multidisciplinary Study*, 595–598. New York: Cambridge University Press.

Schopf, J. W. 1994. Disparate rates, differing fates: Tempo and mode of evolution changed from the Precambrian to the Phanerozoic. *Proceedings of the National Academy of Sciences USA* 91: 6735–6742.

Simpson, G. G. 1944. *Tempo and Mode in Evolution.* New York: Columbia University Press.

Chapter 9

Knoll, A. H., and Holland, H. D. 1995. Oxygen and Proterozoic evolution: An update. In Board on Earth Sciences and Resources, Commission on Geosciences, Environment, and Resources, National Research Council, *Effects of Past Climates on Life*, 21–33. Washington, D.C.: National Academy Press.

Runnegar, R. 1992. Evolution of the earliest animals. In J. W. Schopf, ed., *Major Events in the History of Life*, 65–93. Boston: Jones and Bartlett.

Schopf, J. W. 1992. Times of origin and earliest evidence of major biologic groups. In J. W. Schopf and C. Klein, eds., *The Proterozoic Biosphere: A Multidisciplinary Study*, 587–591. New York: Cambridge University Press.

Schopf, J. W., Haugh, B. N., Molnar, R. E., and Satterthwait, D. F. 1973. On the development of metaphytes and metazoans. *Journal of Paleontology* 47: 1–9.

Summons, R. E., and Walter, M. R. 1990. Molecular fossils and microfossils of prokaryotes and protists from Proterozoic sediments. *American Journal of Science* 290-A: 212–244.

Szathmáry, E., and Maynard Smith, J. 1995. The major evolutionary transitions. *Nature* 374: 227–232.

Chapter 10

Blackmore, V., and Page, A. 1989. *Evolution the Great Debate.* Oxford: Lion Publishing PlC.

Cloud, P. 1983. Early biogeologic history: The emergence of a paradigm. In J. W. Schopf, ed., *Earth's Earliest Biosphere: Its Origin and Evolution*, 14–31. Princeton, N.J.: Princeton University Press.

Goldsmith, D. 1997. *Worlds Unnumbered*. Sausalito, Calif.: University Science Books.

Kuhn, T. 1970. *The Structure of Scientific Revolutions*. 2d ed. Chicago: University of Chicago Press.

Schopf, J. W. 1994. Disparate rates, differing fates: Tempo and mode of evolution changed from the Precambrian to the Phanerozoic. *Proceedings of the National Academy of Sciences USA* 91: 6735–6742.

Chapter 11

Beringer, J.B.A. 1726. *Lithographiae Wirceburgensis*. University of Würzburg, Germany: Mark Anthony Engmann.

Jahn, M. E., and Woolf, D. J. 1963. *The Lying Stones of Dr. Beringer*. Berkeley: University of California Press.

Shen-Miller, J., Mudgett, M. B., Schopf, J. W., Clarke, S., and Berger, R. 1995. Exceptional seed longevity and robust growth, ancient Sacred Lotus from China. *American Journal of Botany* 82: 1367–1380.

Stetter, K. O. 1996. Hyperthermophilic procaryotes. *FEMS Microbiological Review* 18: 149–158.

van Regteren Altena, C. O., and Möckel, J. R. 1967. *Minerals and Fossils in the Teyler Museum*. Haarlem, The Netherlands: Teyler Museum.

Chapter 12

Achenbach, J. 1997. The genesis problem. *Washington Post Magazine*, November 2, 1997, 12–17, 35, 36, 38–40.

Anders, E. 1996. Evaluating the evidence for past life on Mars: Technical comment. *Science* 274: 2119–2121.

Clemett, S. J., and Zare, R. N. 1996. Evaluating the evidence for past life on Mars: Technical comment. *Science* 274: 2122–2123.

Gibson, E. K., Jr., McKay, D. S., Thomas-Keprta, K. L., and Romanek, C. S. 1996. Evaluating the evidence for past life on Mars: Technical comment. *Science* 274: 2125.

Gibson, E. K., Jr., McKay, D. S., Thomas-Keprta, K. L., and Romanek, C. S. 1997. The case for relic life on Mars. *Scientific American* 277: 36–41.

Gladman, B. J., and Burns, J. A. 1996. Mars meteorite transfer: Simulation. *Science* 274: 162.

Goldsmith, D. 1997. *The Hunt for Life on Mars*. New York: Penguin Books.

Kerr, R. A. 1997. Putative martian microbes called microscopy artifacts. *Science* 278: 1706–1707.

McKay, D. S., Gibson, E. K., Jr., Thomas-Keprta, K. L., Vali, H., Romanek, C. S., Clemett, S. J., Chiller, X.D.F., Maechling, C. R., and Zare, R. N. 1996. Search for past life on Mars: Possible relic biogenic activity in Martian meteorite ALH84001. *Science* 273: 924–930.

McKay, D. S., Thomas-Keprta, K. L., Romanek, C. S., Gibson, E. K., and Vali, H. 1996. Search for past life on Mars: Technical comment. *Science* 274: 2124.

Shearer, C. K., and Papike, J. J. 1996. Evaluating the evidence for past life on Mars: Technical comment. *Science* 274: 2121.

Yarus, M. 1997. Is the case persuasive? A skeptical view. *The Planetary Report* 17 (1): 18–19.

Index of Geologic Units and Genera and Species

Subject Index

Abelson, Philip, 57–60; quotation of, 60

acritarchs, 28, 243, 246, 252–259; acanthomorph, 243; etymology of, 243; megasphaeromorph, 28, 252; mesosphaeromorph, 252; Proterozoic extinction of, 253, 255–259; sphaeromorph, 243, 252

Age of: Algae, 34; Dinosaurs, 17; Evident Life, 12; Fish, 17; Flowering Plants, 17; Generalization and Survival, 212; Invertebrates, 17; Microscopic Life, 12; Naked Seed Plants, 17; Seaweeds, 17; Specialization and Extinction, 212; Spore-Producing Plants, 17

Agricola, Georgius, 300

Akberdin Formation microfossils, 220

ALH84001 meteorite, 310–315, 320, 325

animals, 4, 10, 11, 12, 20, 42, 52, 68, 105, 106, 259–263; oldest fossil, 52, 68, 259–263; and Tree of Life, 105, 106

Apex chert microfossils, 75–100; age of, 85–89; bacterial, 96–99; biologic traits of, 94–96; conglomerate containing, 88, 89; cyanobacterium-like, 77, 78, 96–99; geographic source of, 83–86; habitat of, 89–92; preservation of, 76; thin section study of, 77, 92–94

Archean Era, 4, 72, 73, 80, 91, 92, 216; atmosphere during, 91, 92; etymology of, 4; survival of rocks from, 72, 73, 80, 216

atmosphere, 68, 91, 92, 125–128, 170–173, 234, 235; of early Earth, 91, 92, 170–173, 234, 235; evolution of, 68, 170–173; simulated in Miller-type experiments, 125–128

Bacon, Francis, 7

bacteria, 29, 33, 34, 69, 72, 105, 177, 178, 216, 222–224; in Apex chert, 96–99; Gram-negative, 105, 177–181, 216; Gram-positive, 105, 177, 178, 216; mor-

phometrics of, 222–224; Precambrian fossil record of, 216; Walcott's fossil, 29, 33, 34, 69, 268

banded iron formation (BIF), 10, 13, 171–173, plate 6; deposition during "rusting" of the Earth, 171–173

Barberton Mountain Land, South Africa, 80

Barghoorn, Dorothy (Osgood), 63, 64

Barghoorn, Elso S., 26, 42–44, 47, 48, 50, 52, 55–64, 67–69, 71, 76, 264, 265, 267, 269; quotation of, 44, 56, 59, 60

Barghoorn, Frederick, 47

Bauer, Georgius, 300

Belcher Supergoup microfossils, 226

Beringer, Johann Bartholomew Adam, 291–303; quotation of, 297–299, 302

"Beringer's lying stones," 291–303

Bernal, J. D., 123, 266

Berra, Yogi, 273; quotation of, 273

BIF. See banded iron formation

Bitter Springs Formation microfossils, 64–67, 69, 78, 269, 320, 321, plate 1, plate 2

Cambrian Period, 5, 15, 23, 25, 29, 50, 52, 239, 259–263; "explosion" of many-celled animals, 50, 259, 263; fauna of, 23, 29, 52, 67, 259–263; stromatolites from, 25

carbon isotopic signature, 68, 95, 174–181; of photosynthesis, 68, 95, 174–177; of methane-producing archaeans, 179, 181

carbon-14 (^{14}C) dating, 85, 86, 174

Carnegie, Andrew, 25

Carpenter, William B., 20, 21, 265–267

Chaloner, William, 54, 55, 62; quotation of, 54, 55

Chichkan Formation microfossils, plate 2

Chroococcaceae, 215, 216; and modern-fossil look-alikes, 215, 225, 228; Precambrian, 216, 225, 228; survival and growth of, 232–235